T0269045

This volume describes the Pomeron, an object of crucial importance in very high energy particle physics.

The book starts with a general description of the Pomeron within the framework of Regge theory. The emergence of the Pomeron within scalar field theory is discussed next, providing a natural foundation on which to develop the more realistic case of QCD. The reggeization of the gluon is demonstrated and used to build the Pomeron of perturbative QCD. The dynamical nature of the Pomeron is then investigated. The role of the Pomeron in small-x deep inelastic scattering and in diffractive scattering is also examined in detail. The volume concludes with a study of the colour dipole approach to high energy scattering and the explicit role of unitarity corrections.

This book will be of interest to theoretical and experimental particle physicists, and applied mathematicians.

CAMBRIDGE LECTURE NOTES IN PHYSICS 9

General Editors: P. Goddard, J. Yeomans

Quantum Chromodynamics and the Pomeron

CAMBRIDGE LECTURE NOTES IN PHYSICS

Quantum Chromodynamics and the Pomeron

J. R. FORSHAW

University of Manchester

D. A. ROSS

University of Southampton

CAMBRIDGE
UNIVERSITY PRESS

PUBLISHED BY THE PRESS SYNDICATE OF THE UNIVERSITY OF CAMBRIDGE
The Pitt Building, Trumpington Street, Cambridge CB2 1RP United Kingdom

CAMBRIDGE UNIVERSITY PRESS
The Edinburgh Building, Cambridge CB2 2RU, United Kingdom
40 West 20th Street, New York, NY 10011–4211, USA
10 Stamford Road, Oakleigh, Melbourne 3166, Australia

First published 1997

Typeset by the author

A catalogue record for this book is available from the British Library

Library of Congress Cataloguing in Publication data

Forshaw. J. R. (Jeffrey Robert), 1968–
Quantum chromodynamics and the pomeron/J. R. Forshaw, D. A. Ross.
p. cm (Cambridge lecture notes in physics: 9)
Includes bibliographical references and index.
ISBN 521 56880 3 (pbk.)
1. Pomerons.   2. Quantum chromodynamics.   3. Perturbation (Quantum dynamics)
4. Regge theory.   I. Ross, D. A. (Douglas Alan), 1948–
II. Title.   III. Series.
QC793.5.P58F67   1997
539.7–21dc21   97–4006CIP

ISBN 0 521 56880 3 paperback

Transferred to digital printing 2004

Dedicated to the memory of

John Kenneth Storrow
(1943–1996)

Contents

Preface

In recent years, the study of strong interaction physics within the framework of Quantum Chromodynamics (QCD) has largely been restricted to processes which involve a single hard scale (of the order of the centre-of-mass energy). There is a whole wealth of strong interaction physics which is ignored in such a study, including the connection between QCD and Regge theory, which was successfully used to describe certain aspects of the strong interactions before the advent of QCD.

The connection between QCD and Regge theory has attracted much attention in the theoretical community for many years now. Indeed the BFKL equation, which describes what we shall refer to as the perturbative Pomeron, has been known for nearly twenty years. Only recently with the arrival of the HERA and Tevatron colliders has it been possible to perform experiments in the kinematic regime relevant to the perturbative Pomeron. Structure functions at low values of Bjorken x and the observation of rapidity gaps are examples of phenomena which can be used to test the perturbative Pomeron.

The work of those many authors who have contributed to the understanding of the Pomeron in QCD is indeed very formidable. However, to our knowledge, no single self-contained compendium of such work exists. Furthermore many of the papers which have been published on this subject have not been written in a particularly pedagogical style and are therefore not easily understood by a pedestrian reader who wishes learn about the perturbative Pomeron.

In view of the high profile which the Pomeron is now receiving, this lecture note volume is designed to explain the derivation and applications of the perturbative Pomeron from first principles. It is aimed at the level of graduate students who have completed a course in quantum field theory. Certain techniques which may not be included in such a course are briefly reviewed, often in appendices in order not to interrupt the flow of the text. It is, of course, also hoped that more senior physicists who wish to become familiar with the perturbative Pomeron will find this volume useful.

Almost all of the material in this volume is the work of other authors and only rarely have we alluded to our own modest contributions to this subject. References have been given to papers which are specifically relevant to topics covered in the text and these are by no means intended to form a complete bibliography of the vast number of papers that have been published in this field.

We begin with a review of the Pomeron in the old Regge theory, largely for the benefit of the, by now, majority of physicists who are too young to have met such material in graduate school (this includes one but not both of the authors). We then present a toy model example which introduces the reader to the techniques that are used to derive the perturbative Pomeron in QCD. One of the essential ingredients in the BFKL approach to the perturbative Pomeron is the concept of the reggeized gluon. The demonstration that the gluon does reggeize is given in Chapter 3. It is a necessarily involved demonstration. The reader who is prepared to take the result on trust may wish to skip from section 3.2 to the end of the chapter and proceed to Chapter 4, where the BFKL equation is derived. In Chapters 6 and 7 we discuss applications of the perturbative Pomeron to processes which are currently under experimental observation at the HERA and Tevatron colliders. We end the volume with a discussion of recent progress that has been made on the restoration of unitarity at very high energy.

This book has its own page on the World Wide Web at the URL "http : //h2.ph.man.ac.uk/~forshaw/book.html". The page includes a list of misprints and corrections and we would appreciate communications reporting additional errors.

J.R. Forshaw
D.A. Ross
February 1997

Acknowledgement

Before and during the preparation of this volume our understanding of both theoretical and experimental aspects of this subject has been greatly enhanced by useful and enjoyable conver-
sations with Halina Abramowicz, Mike Albrow, Kevin Anderson,
Jochen Bartels, Jon Butterworth, Mandy Cooper-Sarkar, Stefano
Catani, Frank Close, Jean-René Cudell, John Dainton, Robin De-
venish, Sandy Donnachie, John Ellis, Keith Ellis, Norman Evan-
son, Brian Foster, Lonya Frankfurt, Robert Hancock, Peter Harri-
man, Francesco Hautmann, Jan Kwiecinski, Mark Lancaster, Peter Landshoff, Genya Levin, Lev Lipatov, Hans Lotter, Norman
McCubbin, Martin McDermott, Uri Maor, Pino Marchesini, Alan
Martin, Andy Mehta, Al Mueller, Basrab Nicolescu, Kolya Niko-
laev, Julian Phillips, Dick Roberts, Graham Ross, Misha Ryskin,
Gavin Salam, Graham Shaw, Dave Soper, Mike Sotiropoulos,
John Storrow, Peter Sutton, Robert Thorne, Bryan Webber, Alan
White and Mark Wüsthoff.

We are particularly grateful to Peter Landshoff, Gavin Salam
and Peter Sutton for permission to reproduce some of their figures,
and to Sandy Donnachie and Mike Sotiropoulos for allowing us the
use of their notes, which inspired some of the material in the first
chapter.

Last but by no means least we are grateful to Andrea and Jackie
for their unending patience and support.

1
What is a Pomeron?

Before the advent of the field theoretic approach (QCD), a good deal of progress had already been made in developing an understanding of the scattering of strongly interacting particles. This progress was founded on some very general properties of the scattering matrix. Regge theory provided a natural framework in which to discuss the scattering of particles at high centre-of-mass energies.

With the arrival of QCD much attention was diverted away from the 'old fashioned' approach to the strong interactions. Interest was re-ignited within the particle physics community with the arrival of colliders capable of delivering very large centre-of-mass energies (e.g. the HERA collider at DESY and the Tevatron collider at FNAL). For the first time physicists started to investigate in earnest the properties of QCD at high energies and compare them with the predictions of the Regge theory.

The high energy limit provides the arena in which the Regge properties of QCD can be studied. It is the meeting place of the 'old' particle physics with the 'new'. Since by 'old' we mean over 30 years ago it is necessary to commence our study of high energy scattering in QCD with an introduction to (or recap of) Regge theory. This chapter will contain a 'whistle-stop tour' of Regge theory and Pomeron phenomenology. We keep this to the minimum which will be required in order to follow the subsequent chapters and refer the interested reader to the literature (e.g. Collins (1977)) for further details.

1.1 Life before QCD

Before the development of QCD nobody dared to apply quantum field theory to the strong interactions. Instead, physicists tried

1

to extract as much as possible by studying the consequences of a (reasonable) set of postulates about the S-matrix, whose abth element is the overlap between the *in*-state (free particles state as $t \to -\infty$), $|\,a\rangle$, and the *out*-state (free particles state as $t \to +\infty$), $|\,b\rangle$,

$$S_{ab} = \langle b_{out} \,|\, a_{in}\rangle.$$

Postulate 1:
The S-matrix is Lorentz invariant.

This means that it can be expressed as a function of the (Lorentz invariant) scalar products of the incoming and outgoing momenta. For two-particle to two-particle scattering,

$$a + b \to c + d,$$

these are most effectively described in terms of the Mandelstam variables, s, t, and u defined by

$$\begin{aligned}
s &= (p_a + p_b)^2 \\
t &= (p_a - p_c)^2 \\
u &= (p_a - p_d)^2,
\end{aligned}$$

as well as the four masses, m_a, m_b, m_c, m_d. The total energy of the system in the centre-of-mass frame is $\sqrt{s}$ and t is the square of the four-momentum exchanged between particles a and c and is related to the scattering angle. u is not an independent variable since by conservation of momentum we can show that[†]

$$s + t + u = m_a^2 + m_b^2 + m_c^2 + m_d^2.$$

We therefore write a two-particle to two-particle scattering amplitude as $\mathcal{A}(s,t)$, a function of s and t only (the amplitude also depends on the masses of the external particles).

For two-particle to n particle scattering processes there are $3n - 4$ independent invariants.

Postulate 2:
The S-matrix is unitary:

$$SS^\dagger = S^\dagger S = \mathbb{1}.$$

[†] Throughout this book we work in the system of units $\hbar = c = 1$.

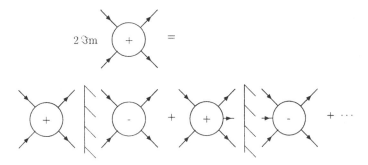

Fig. 1.1. The Cutkosky rules for a two-particle to two-particle amplitude. The shaded cut line denotes that the intermediate particles are on mass-shell whilst the + and − signs denote the amplitude and its hermitian conjugate respectively.

This is a statement of conservation of probability, i.e. the probability for an *in*-state to end up in a particular *out*-state, summed over all possible *out*-states, must be unity.

The scattering amplitude, $\mathcal{A}_{ab}$, for scattering from an *in*-state $|a\rangle$ to an *out*-state $|b\rangle$ is related to the S-matrix element by

$$S_{ab} = \delta_{ab} + i(2\pi)^4\delta^4 \left(\sum_a p_a - \sum_b p_b\right) \mathcal{A}_{ab}$$

$((2\pi)^4\delta^4 \left(\sum_a p_a - \sum_b p_b\right) \mathcal{A}_{ab}$ is often called the T-matrix element T_{ab} where $S = \mathbb{1} + iT$) and the unitarity of the S-matrix leads to the relation

$$2\Im\text{m}\, \mathcal{A}_{ab} = (2\pi)^4\delta^4 \left(\sum_a p_a - \sum_b p_b\right) \sum_c \mathcal{A}_{ac}\mathcal{A}_{cb}^\dagger. \qquad (1.1)$$

This gives us the Cutkosky (1960) rules, which allow us to determine the imaginary part of an amplitude by considering the scattering amplitudes of the incoming and outgoing states into all possible 'intermediate' states. These rules will be used extensively in later chapters. For the case of two-particle to two-particle scattering the Cutkosky rules are shown schematically in Fig. 1.1. Here the shaded 'cut' line means that the intermediate particles are taken to be on their mass-shell and an integral is performed over the phase space of the intermediate particles. The minus signs in the amplitudes on the right of the cuts mean that the hermitian conjugate is taken, i.e. the *in-* and *out-*states are interchanged and the complex conjugate is taken (in perturbation theory this implies that the sign of the $i\epsilon$ for each internal propagator is reversed).

An important special case of the Cutkosky rules is the **optical theorem**, which relates the imaginary part of the forward elastic amplitude, $\mathcal{A}_{aa}$, to the total cross-section for the scattering of the (two-particle) state, $|a\rangle$,

$$2\Im\text{m}\mathcal{A}_{aa}(s,0) = (2\pi)^4 \sum_n \delta^4 \left(\sum_f p_f - \sum_a p_a \right) |\mathcal{A}_{a-n}|^2 = F\,\sigma_{\text{tot}},$$

(1.2)

where F is the flux factor (for $\sqrt{s}$ much larger than the masses of the incoming particles $F \approx 2s$).

Postulate 3:

The S-matrix is an analytic function of Lorentz invariants (regarded as complex variables), with only those singularities required by unitarity.

It can be shown that this 'analyticity' property is a consequence of causality, i.e. that two regions with a space-like separation do not influence each other.

Analyticity has a number of important and useful consequences. Combined with unitarity we are able to establish the existence of the s-plane singularity structure of the amplitude $\mathcal{A}(s,t)$ shown in Fig. 1.2, i.e. there are s-plane cuts with branch points corresponding to physical thresholds. These arise because the n-particle states must contribute to the imaginary part of the amplitude if s is greater than the n-particle threshold (see Eq.(1.1)). The imaginary part of the amplitude is

$$\Im\text{m}A(s,t) = \frac{A(s,t) - A(s,t)^*}{2i}.$$

Below threshold there are no contributions to the imaginary part and so there exists a region on the real s-axis (around the origin) where the amplitude is purely real. This means that we can use the Schwarz reflection principle, which states that a function (of s) which is real on some part of the real s-axis satisfies

$$\mathcal{A}(s,t)^* = \mathcal{A}(s^*,t)$$

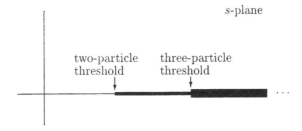

Fig. 1.2. The cuts on the positive real axis in the complex s-plane.

throughout its domain of analyticity. So, in order to have an imaginary part for real s above threshold, we need a cut along the real axis with branch point at the threshold energy.[†] Using the Schwarz reflection principle we can write

$$\Im m A(s + i\epsilon, t) = \frac{A(s + i\epsilon, t) - A(s - i\epsilon, t)}{2i}$$

in the region where the amplitude is analytic, e.g. for real s and ϵ. This is non-zero for real s above threshold and allows us to define the imaginary part of the physical scattering amplitude above threshold as

$$\Im m \, \mathcal{A}(s, t) = \frac{1}{2i} \lim_{\epsilon \to 0} [\mathcal{A}(s + i\epsilon, t) - \mathcal{A}(s - i\epsilon, t)]. \qquad (1.3)$$

The right hand side is called the s-**channel discontinuity** and is often denoted by $\Delta_s \mathcal{A}(s, t)$.[‡] Analyticity also implies, as we shall shortly show, that there are cuts along the negative real axis.

A further consequence of analyticity is crossing symmetry. Consider the scattering process

$$a + b \to c + d, \qquad (1.4)$$

and write its amplitude as $\mathcal{A}_{a+b\to c+d}(s, t, u)$ (we have reinstated the variable u for the sake of symmetry but understand that this is *not* an independent variable). Now in the physical kinematic regime for the process (1.4) we have $s > 0$ and $t, u < 0$. Since the amplitude is an analytic function it may be analytically continued

---

[†] This is true for ≥ 2 particles in the intermediate state. For single particle production, i.e. a bound state of mass m, we have a pole at $s = m^2$.

[‡] This corresponds to the definition of the physical scattering amplitude as the limit $\lim_{\epsilon \to 0} A(s + i\epsilon, t)$.

to the region $t > 0$ and $s, u < 0$. This gives the amplitude for the t-channel process,

$$a + \bar{c} \to \bar{b} + d, \qquad (1.5)$$

where $\bar{b}$, $\bar{c}$ mean the antiparticles of particles b and c respectively. Thus we have

$$\mathcal{A}_{a+\bar{c}\to\bar{b}+d}(s, t, u) = \mathcal{A}_{a+b\to c+d}(t, s, u) \qquad (1.6)$$

and similarly for the u-channel process,

$$a + \bar{d} \to \bar{b} + c, \qquad (1.7)$$

we have

$$\mathcal{A}_{a+\bar{d}\to\bar{b}+c}(s, t, u) = \mathcal{A}_{a+b\to c+d}(u, t, s). \qquad (1.8)$$

Since the amplitude for the t-channel and u-channel processes also have imaginary parts and consequently physical thresholds, there must be cuts along the real positive t and u axes with branch points at these thresholds. Now $u = \sum_i m_i^2 - s - t$, so that the existence of a threshold at $u = u_{th}$ for positive u (at fixed t) means that as well as a branch point at positive $s = s_{th}^+$ corresponding to the physical threshold for the s-channel process, the amplitude, $\mathcal{A}(s, t)$ must have a cut along the negative real s-axis with a branch point at $s = s_{th}^- = \sum_i m_i^2 - t - u_{th}$.

The next important consequence of analyticity which we shall make use of is that it enables us to reconstruct the real part of an amplitude from its imaginary part using dispersion relations. We refer to the standard texts on mathematical physics for those readers unfamiliar with dispersion relations (e.g. Mathews & Walker (1970)).

The Cauchy integral formula allows us to write

$$\mathcal{A}(s, t) = \frac{1}{2\pi i} \oint_c \frac{\mathcal{A}(s', t)}{(s' - s)} \, ds',$$

where C is a contour that does not enclose any of the singularities of $\mathcal{A}$. Such a contour is shown in Fig. 1.3. It goes around the cuts along the positive and negative real axes and around the semicircles at infinity. The contributions to the contour integral from

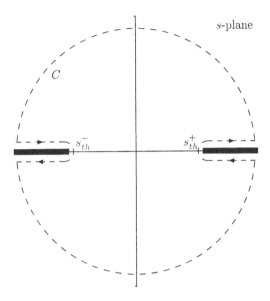

Fig. 1.3

the parts that surround the cuts are

$$\int_{s_{th}^+}^{\infty} ds' \frac{A(s'+i\epsilon,t)}{(s'-s)} + \int_{\infty}^{s_{th}^+} ds' \frac{A(s'-i\epsilon,t)}{(s'-s)}$$

$$+ \int_{-\infty}^{s_{th}^-} ds' \frac{A(s'+i\epsilon,t)}{(s'-s)} + \int_{s_{th}^-}^{-\infty} ds' \frac{A(s'-i\epsilon,t)}{(s'-s)}.$$

Provided $A(s,t)$ falls to zero as $|s| \to \infty$, the contribution to the contour integration from the semi-circles at infinity may be neglected and using Eq.(1.3) we end up with the dispersion relation[†]

$$A(s,t) = \frac{1}{\pi} \int_{s_{th}^+}^{\infty} \frac{\Im m A(s',t)}{(s'-s)} ds' + \frac{1}{\pi} \int_{-\infty}^{s_{th}^-} \frac{\Im m A(s',t)}{(s'-s)} ds'. \quad (1.9)$$

In the second of these integrals the imaginary part of the amplitude for $s < s_{th}^-$ is obtained from the Cutkosky rules applied to the u-channel process (1.7), i.e.

$$\Im m A(s < s_{th}^-, t) = -\Delta_u A(s,t).$$

[†] We have assumed no contribution from bound state poles which generally add extra contributions.

If the amplitude does *not* vanish as $|s| \to \infty$, then we have to make subtractions, i.e. we divide the amplitude by as many factors of $s - s_i$ as are necessary to produce a vanishing contribution from the semi-circles at infinity (the s_i are arbitrary and define the points at which the subtractions take place). For example, making one subtraction at $s = s_0$ we obtain the subtracted dispersion relation

$$
\mathcal{A}(s,t) = \mathcal{A}(s_0,t) + \frac{(s-s_0)}{\pi} \int_{s_{th}^+}^{\infty} \frac{\Im m \mathcal{A}(s',t)}{(s'-s)(s'-s_0)} ds'
$$
$$
+ \frac{(s-s_0)}{\pi} \int_{-\infty}^{s_{th}^-} \frac{\Im m \mathcal{A}(s',t)}{(s'-s)(s'-s_0)} ds'. \qquad (1.10)
$$

For our purposes we shall require the subtracted dispersion relation which allows us to reconstruct a function of s whose imaginary part is given by

$$
A \left(\ln s \right)^n .
$$

Equation (1.10) allows us to establish that, to leading order in $\ln s$, this function is purely real and equal to

$$
- \frac{A}{(n+1)\pi} \left(\ln s \right)^{n+1} ,
$$

where we have used Eq.(1.3) to write

$$
\ln \left(-s \right) = \ln \left(s \right) - i\pi .
$$

Thus we see how, from three rather general postulates coupled with the spectrum of elementary particles, we can develop at least a set of self-consistency conditions for amplitudes and their relation to each other. Unitarity relates the imaginary parts of amplitudes to sums of products of other amplitudes, and dispersion relations then allow us to determine the corresponding real parts. The application of this process is called **a bootstrap** and it does not make any assumption about any underlying quantum field theory which may describe the dynamics of the strong interactions.

A further ingredient needed for the bootstrap is the asymptotic behaviour of amplitudes. Once we know these and their analytic structure then analyticity can be used to reconstruct the amplitudes. Determination of the asymptotic behaviour of amplitudes is the goal of Regge theory (Regge (1959, 1960)).

1.2 Sommerfeld–Watson transform

Let us consider a two-particle to two-particle scattering process in the t-channel (Eq.(1.5)) at a centre-of-mass energy, $\sqrt{s}$, which is much larger than the masses of the external particles. The amplitude can be expanded as a series in Legendre polynomials, $P_l(\cos\theta)$, where θ is the (centre-of-mass frame) scattering angle and is related to s, t by

$$\cos\theta = 1 + \frac{2t}{s}.$$

This expansion is called the **partial wave expansion**, namely,

$$A_{a\bar{c}\rightarrow\bar{b}d}(s,t) = \sum_{l=0}^{\infty}(2l+1)\,a_l(s)\,P_l\left(1+2t/s\right). \qquad (1.11)$$

$P_l(z)$ is a polynomial in z of degree l, and the functions $a_l(s)$ are called the **partial wave amplitudes**.

From the property of crossing symmetry (Eq.(1.6)) this may be continued into the s-channel by interchanging s and t to give

$$A_{ab\rightarrow cd}(s,t) = \sum_{l=0}^{\infty}(2l+1)\,a_l(t)\,P_l\left(1+2s/t\right). \qquad (1.12)$$

Sommerfeld (1949), following Watson (1918), rewrote this partial wave expansion in terms of a contour integral in the complex angular momentum (l) plane as

$$A(s,t) = \frac{1}{2i}\oint_C dl(2l+1)\frac{a(l,t)}{\sin\pi l}P(l,1+2s/t), \qquad (1.13)$$

where the contour C surrounds the positive real axis as shown in Fig. 1.4. The Legendre polynomials can be expressed in terms of hypergeometric functions and analytically continued in l, giving the analytic function $P(l,z)$. The function $a(l,t)$ is an analytic continuation of the partial wave amplitudes $a_l(t)$. The denominator $\sin\pi l$ vanishes for integer l giving rise to poles which then reproduce Eq.(1.12).

1.3 Signature

It is now natural to ask if the function $a(l,t)$ is unique. At first sight it appears that it is not. For example we could add to $a(l,t)$ any analytic function which vanishes at integer values of

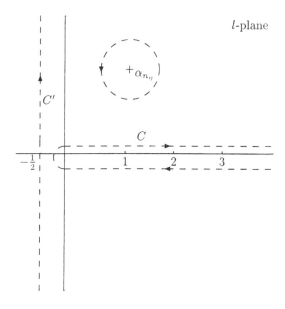

Fig. 1.4. Sommerfeld–Watson transform.

l without affecting the above result. However, using a theorem by Carlson (1914), it can be shown that $a(l,t)$ is unique provided $a(l,t) < \exp(\pi|l|)$ as $|l| \to \infty$. Unfortunately there are contributions to the partial wave amplitudes which alternate in sign, i.e. are proportional to $(-1)^l$ and so the required inequality is violated along the imaginary axis. It is therefore necessary to introduce *two* analytic functions $a^{(+1)}(l,t)$ and $a^{(-1)}(l,t)$ which are the analytic continuations of the even and odd partial wave amplitudes. Thus we have

$$A(s,t) = \frac{1}{2i} \oint_C dl \frac{(2l+1)}{\sin \pi l} \sum_{\eta=\pm 1} \frac{(\eta + e^{-i\pi l})}{2} a^{(\eta)}(l,t) \, P(l, 1 + 2s/t),$$

$$(1.14)$$

where η, which takes the values ± 1, is called the **signature** of the partial wave and $a^{(+1)}(l,t)$ and $a^{(-1)}(l,t)$ are called the even- and odd-signature partial wave functions. The prefactors $\frac{1}{2}(\eta + \exp(-i\pi l))$ are called the signature factors.

1.4 Regge poles

The next step is to deform the contour C of Fig. 1.4 to the contour C', which runs parallel to the imaginary axis with $\Re\, l = -\frac{1}{2}$. In order to do this we must encircle any poles or cuts that the functions $a^{(\eta)}(l,t)$ may have at $l = \alpha_{n_\eta}(t)$ and pick up $2\pi i \times$ the residue of that pole. For the particular case of simple poles only we arrive at

$$
\mathcal{A}(s,t) = \frac{1}{2i} \int_{-\frac{1}{2}-i\infty}^{-\frac{1}{2}+i\infty} dl \left[\frac{(2l+1)}{\sin \pi l} \sum_{\eta=\pm 1} \frac{(\eta + e^{-i\pi l})}{2} a^{(\eta)}(l,t) \right.
$$

$$
\left. \times\, P(l, 1+2s/t) \right]
$$

$$
+ \sum_{\eta=\pm 1} \sum_{n_\eta} \frac{(\eta + e^{-i\pi \alpha_{n_\eta}(t)})}{2} \frac{\tilde{\beta}_{n_\eta}(t)}{\sin \pi \alpha_{n_\eta}(t)} P(\alpha_{n_\eta}(t), 1+2s/t). \quad (1.15)
$$

The simple poles $\alpha_{n_\eta}(t)$ are called even- ($\eta = +1$) and odd- ($\eta = -1$) signature **Regge poles** and $\tilde{\beta}_{n_\eta}(t)$ are the residues of the poles multiplied by $\pi(2\alpha_{n_\eta}(t)+1)$.

Throughout this book we shall be concerned with the **Regge region**, i.e. $s \gg |t|$. In this limit the Legendre polynomial is dominated by its leading term and we have

$$
P_l(1 + 2s/t) \xrightarrow{s \gg |t|} \frac{\Gamma(2l+1)}{\Gamma^2(l+1)} \left(\frac{s}{2t} \right)^l,
$$

where $\Gamma(x)$ is the Euler gamma function. In this limit the contribution to the right hand side of Eq.(1.15) from the integral along the contour C' vanishes as $s \to \infty$, so that it may be neglected. It should now be clear why we exploited the crossing symmetry to write Eq.(1.12) and why we deformed the contour as in Fig. 1.4 – we wanted to exploit the asymptotic behaviour of the Legendre polynomial so as to isolate the high energy behaviour of the scattering amplitude in the Regge region. In fact we need only consider the contribution from the Regge pole with the largest value of the real part of $\alpha_{n_\eta}(t)$ (the leading Regge pole). Thus we have

$$
\mathcal{A}(s,t) \xrightarrow{s \to \infty} \frac{(\eta + e^{-i\pi \alpha(t)})}{2} \beta(t) s^{\alpha(t)}, \quad (1.16)
$$

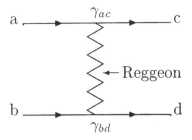

Fig. 1.5. A Regge exchange diagram.

where $\alpha(t)$ is the position of the leading Regge pole (at some value of t) and η is its signature. Some factors depending on t (but *not* on s) have been absorbed into the function $\beta(t)$.

Although we have assumed only simple poles in arriving at Eq.(1.16) it is possible that there are also non-simple poles and cuts which would lead to additional contributions to the amplitudes. We shall show that whereas the simple pole model works well for certain hadronic processes, leading logarithm perturbation theory can in general give rise to cuts.

1.5 Factorization

We can view the amplitude given by Eq.(1.16) as the exchange in the t-channel of an object with 'angular momentum' equal to $\alpha(t)$. This is of course not a particle since the 'angular momentum' is not integer (or half-integer) and it is a function of t. It is called a **Reggeon**. We can view a Reggeon exchange amplitude as the superposition of amplitudes for the exchanges of all possible particles in the t-channel. The amplitude can be factorized as shown in Fig. 1.5 into a coupling $\gamma_{ac}(t)$ of the Reggeon between particles a and c, a similar coupling $\gamma_{bd}(t)$ between particles b and d and a *universal* contribution from the Reggeon exchange. The couplings γ are functions of t only. Thus we obtain

$$\mathcal{A}(s,t) \xrightarrow{s \to \infty} \frac{(\eta + e^{-i\pi\alpha(t)})}{2 \sin \pi\alpha(t)} \frac{\gamma_{ac}(t)\gamma_{bd}(t)}{\Gamma(\alpha(t))} s^{\alpha(t)}. \tag{1.17}$$

We have explicitly extracted a factor of $\Gamma(\alpha(t))$ in defining the couplings γ. The reason for this is that if $\alpha(t)$ takes an integer value for some value of t then the amplitude has a pole. For positive integers this can be understood as the exchange (in the t-channel) of a resonance particle with integer spin, but we would not expect such resonances with negative values of 'spin' . Such poles are called nonsense poles, and are cancelled by the factor $1/\Gamma\left(\alpha(t)\right)$, which has zeroes at $\alpha(t) = 0, -1, -2 \cdots$.

One immediate consequence of Eq.(1.17) is the relation between the ρ-parameter, defined to be

$$\rho = \frac{\Re e\, \mathcal{A}}{\Im m\, \mathcal{A}}, \qquad (1.18)$$

and the signature and position of the (leading) Regge pole. The couplings $\gamma_{ac}(t)$ and $\gamma_{bd}(t)$ are expected to be real functions of t and so from Eqs.(1.17) and (1.18) we have

$$\rho = -\frac{\eta + \cos \pi\alpha(t)}{\sin \pi\alpha(t)}. \qquad (1.19)$$

1.6 Regge trajectories

If we consider the t-channel process, (1.5), with t positive we expect the amplitude to have poles corresponding to the exchange of physical particles of spin, J_i, and mass m_i, where $\alpha(m_i^2) = J_i$.

Chew & Frautschi (1961, 1962) plotted the spins of low lying mesons against square mass and noticed that they lie in a straight line as shown in Fig. 1.6. In other words $\alpha(t)$ is a linear function of t,

$$\alpha(t) = \alpha(0) + \alpha' t$$

(at least for positive t). From Fig. 1.6 we obtain the values

$$\begin{aligned} \alpha(0) &= 0.55 \\ \alpha' &= 0.86 \ \text{GeV}^{-2}. \end{aligned} \qquad (1.20)$$

We shall see that this linearity continues for negative values of t.

From the s-dependence of the amplitude given in Eq.(1.17) we can deduce that the asymptotic s-dependence of the differential cross-section is given (for a linear trajectory) by

$$\frac{d\sigma}{dt} \propto s^{(2\alpha(0)-2\alpha't-2)}. \qquad (1.21)$$

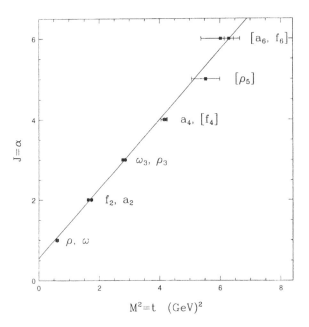

Fig. 1.6. The Chew–Frautschi plot.

If we consider a process in which isospin, $I = 1$, is exchanged in the t-channel, such as

$$\pi^- + p \rightarrow \pi^0 + n, \qquad (1.22)$$

then we expect the Regge trajectory which determines the asymptotic s-dependence to be the one containing the $I = 1$ even parity mesons (the ρ-trajectory). Inserting the values Eq.(1.20) into Eq.(1.21) gives a very good fit to data over a wide range (20–200 GeV) of pion energies, as can be seen in Fig. 1.7.

The Regge trajectory has a further interesting feature. At $t = -0.64$ GeV^2 the trajectory passes through zero. This is an example of a nonsense pole (there cannot be a resonance with negative square mass) and, as explained above, it must decouple from the amplitude. The distinct dip observed in the differential cross-section for the process (1.22) plotted in Fig. 1.8 could well be evidence for the decoupling of this nonsense pole.

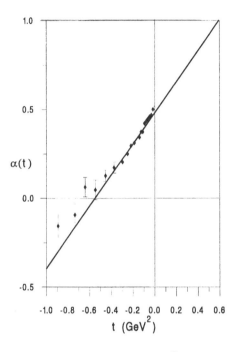

Fig. 1.7. $\alpha(t)$ obtained from $\pi^- p \to \pi^0 n$ data in the pion energy range 20.8–199.3 GeV by Barnes *et al.* (1976). The straight line is obtained by extrapolating the trajectory of Fig. 1.6 (see Eq.(1.20)).

1.7 The Pomeron

From the intercept of the Regge trajectory which dominates a particular scattering process and the optical theorem (Eq.(1.2)) we can obtain the asymptotic behaviour of the total cross-section for that process, namely,

$$\sigma_{\text{tot}} \propto s^{(\alpha(0)-1)}. \qquad (1.23)$$

For the ρ-trajectory considered in the last section $\alpha(0) < 1$, which means that the cross-section for a process with $I = 1$ exchange falls as s increases.

Pomeranchuk (1956) and Okun & Pomeranchuk (1956) proved from general assumptions that in any scattering process in which

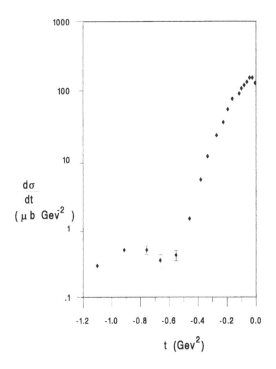

Fig. 1.8. Data on $d\sigma/dt$ for the process $\pi^- p \rightarrow \pi^0 n$ at a beam energy of 20.8 GeV from Barnes *et al.* (1976). The differential cross-section has a dip at $t \approx -0.6$ GeV^2.

there is charge exchange the cross-section vanishes asymptotically (the Pomeranchuk theorem). Foldy & Peierls (1963) noticed the converse, namely, that if for a particular scattering process the cross-section does *not* fall as s increases then that process must be dominated by the exchange of vacuum quantum numbers (i.e. isospin zero and even under the operation of charge conjugation).

It is observed experimentally that total cross-sections do not vanish asymptotically. In fact they rise slowly as s increases. If we are to attribute this rise to the exchange of a single Regge pole then it follows that the exchange is that of a Reggeon whose intercept, $\alpha_P(0)$, is greater than 1, and which carries the quantum

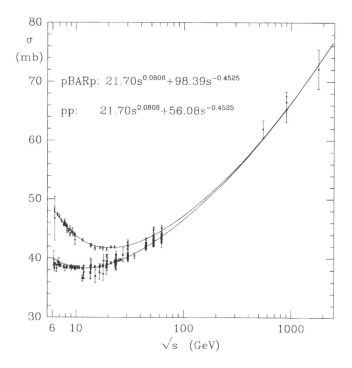

Fig. 1.9. Data for p–p and p–$\bar{p}$ total cross-sections and the fit of Eq.(1.24).

numbers of the vacuum.[†] This trajectory is called the **Pomeron** and is named after its inventor Pomeranchuk (1958).

Unlike the Regge trajectory of Fig. 1.6 the physical particles which would provide the resonances for integer values of $\alpha_P(t)$ for positive t have not been conclusively identified. Particles with the quantum numbers of the vacuum are difficult to detect, but such particles can exist in QCD as bound states of gluons (glueballs).

1.8 Total cross-sections

Fig. 1.9 shows a compilation of data for the total cross-sections

---

[†] The particles with $I = 0$ shown on the trajectory in Fig. 1.6 do *not* have the quantum numbers of the vacuum since they are odd under charge conjugation.

for proton–proton (p–p) and proton–antiproton (p–$\bar{p}$) scattering,
together with a fit due to Donnachie & Landshoff (1992):

$$\sigma_{pp} = 21.7\, s^{0.08} + 56.1\, s^{-0.45}\ \text{mb}$$
$$\sigma_{\bar{p}p} = 21.7\, s^{0.08} + 98.4\, s^{-0.45}\ \text{mb} \qquad (1.24)$$

(with s in GeV^2). These parameters were determined *before*
the measurement of the p–p cross-section at the Fermilab Tevatron accelerator (from fitting to a wide range of data below
$\sqrt{s} = 100$ GeV).

The first term on the right hand side of Eq.(1.24) is the Pomeron
contribution and it is common to both p–p and p–$\bar{p}$ cross-sections,
coupling with the same strength to the proton and antiproton because the Pomeron carries the quantum numbers of the vacuum
and therefore cannot distinguish between particles and antiparticles. The second term, on the other hand, is a sub-leading term
which is due to the exchange of a Regge trajectory with intercept
0.55 (the intercept of the Regge trajectory shown in Fig. 1.6) and
this trajectory can (and does) have different couplings to particles and antiparticles. This accounts for the difference between the
p–p and p–$\bar{p}$ cross-sections at low s (this difference vanishes as s
increases by the Pomeranchuk theorem).

This fit tells us that the Pomeron has intercept $\alpha_P(0) = 1.08$.
This is slightly above 1 and will eventually lead to a violation of
the bound derived by Froissart (1961) and Martin (1963) which is
derived using unitarity and the partial wave expansion (we present
a physical argument for the Froissart–Martin bound in Chapter 8).
They showed that, as s tends to infinity, total hadronic cross-
sections must satisfy the inequality

$$\sigma_{\text{tot}} < A \ln^2 s, \qquad (1.25)$$

where the constant A is determined by the pion mass and is expected to be ~ 60 mb. However, since the intercept is only very
slightly above 1, this violation does not occur for momenta lower
than the Planck scale! It is not unreasonable that physics beyond the exchange of the single Pomeron pole enters to ensure
the ultimate preservation of unitarity (in fact it is known that
multiple Pomeron exchanges are able to tame the asymptotic rise
of the cross-section). Another point of view is to argue that the
intercept of 1.08 is only an effective intercept and that the underlying mechanism which gives rise to it is not the result of single

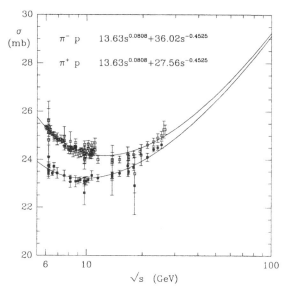

Fig. 1.10. Total cross-sections for $\pi^+\!-\!p$ and $\pi^-\!-\!p$ scattering.

Pomeron exchange but has contributions from the exchange of two or more Pomerons (so-called **Regge cuts**).

If the high energy behaviour of the total cross-section is indeed a result of the superposition of the two Regge exchanges, with intercepts as quantified in Eq.(1.24), then since the intercepts are universal we expect them to be able to describe other total cross-sections. This is indeed the case, as can be seen from Fig. 1.10 for the case of pion–proton scattering and Fig. 1.11 for (on-shell) photon–proton scattering.

1.9 Differential elastic cross-sections

In order to determine the slope, α'_P, of the Pomeron trajectory it is necessary to consider the differential cross-section, e.g. for elastic $p\!-\!p$ or $p\!-\!\bar{p}$ scattering, over a range of s and at different values of t. A collection of data ranging from ISR at CERN to the Tevatron at Fermilab give a good fit to a linear Pomeron trajectory with

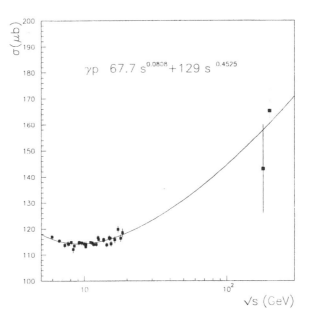

Fig. 1.11. The cross-section for γ–p scattering.

slope

$$\alpha'_P = 0.25 \text{ GeV}^{-2}.$$

From this slope we can determine that $\alpha_P(t)$ reaches the value 2 at $t = 3.7$ GeV^2 and we should expect a spin two particle with mass $\sqrt{3.7} = 1.9$ GeV and the quantum numbers of the vacuum. The WA91 collaboration at CERN (Abatzis *et al.* (1994)) has announced evidence for a candidate glueball state with this mass. This could well be the first observed particle to lie on the Pomeron trajectory.

The couplings, $\gamma(t)$, of the Pomeron can also be obtained from the t-dependence of differential elastic cross-sections (at fixed s). It turns out that the data are well fitted by taking the Pomeron coupling $\gamma(t)$ to be proportional to the electromagnetic form factor of the hadron to which the Pomeron couples. In other words the Pomeron couples to hadrons in the same way as the photon. Thus when the Pomeron couples to hadrons it appears to behave like a point particle. One immediate consequence of this, as was noted by Landshoff & Polkinghorne (1971), is the quark-counting rule which tells us that the Pomeron couples to one constituent quark at a time inside a hadron, so that the coupling to that hadron is expected to be proportional to the number of valence quarks. This quark-counting rule is well supported by the fact that the coefficients of the Pomeron term in the fits to p–p and π–p scattering are in the ratio 1.6:1, which is just slightly higher than the ratio 3:2 that would be expected from the fact that the proton has three valence quarks whereas the pion has only two.

The ρ-parameter (Eq.(1.18)) can also be obtained from the differential elastic cross-section at zero momentum transfer and the total cross-section. The former is proportional to the sum of the squares of the real and imaginary parts of the scattering amplitude, whereas the latter is related by the optical theorem to the imaginary part of the amplitude. Thus we have

$$\frac{d\sigma^{\text{el}}(s,0)}{dt} = \frac{(1+\rho^2)}{16\pi}|\sigma_{\text{tot}}|^2. \tag{1.26}$$

Experimental values such as those of Augier *et al.* (1993) from the UA4 collaboration at CERN give a value of ρ of about 0.1 at $\sqrt{s} \approx 100$ GeV. In other words the amplitude for Pomeron exchange is dominated by its imaginary part. From the fact that the intercept of the Pomeron is close to 1 and Eq.(1.19) we can deduce that the Pomeron must have even signature ($\eta = 1$).

1.10 Diffractive dissociation

At sufficiently high energies elastic-scattering events are rather difficult to detect since the particles scatter through small angles. However, the Pomeron enters into several other processes. One of these is the process of diffractive dissociation in which one of the incident particles remains unchanged and just scatters through a small angle, but the other incident particle receives enough energy for it to break up into its constituent partons, which then hadronize to produce clusters of hadrons.

It is convenient to view such a process from the point of view suggested by Fig. 1.12, where a Pomeron is 'emitted' from the 'parent' hadron (with momentum p_2 and which remains intact after the scattering) with some fraction ξ of its momentum. The

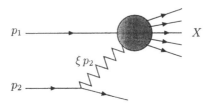

Fig. 1.12. A diffractive dissociation process in which the exchanged Pomeron carries a fraction ξ of the momentum p_2 of one of the incoming hadrons.

upper vertex can be thought of as 'hadron–Pomeron' scattering producing some final hadronic state, X.

Such events have a large **rapidity gap** between the 'parent' hadron and the hadrons in the hadronic system, X. The rapidity, y_i, of particle i is defined as

$$y_i = \frac{1}{2}\ln\left(\frac{E_i + p_{zi}}{E_i - p_{zi}}\right),$$

where the z-axis is taken along the incident beam direction. Since the scattering angle is small ($|t|$ is much smaller than s) the 'parent' hadron emerges almost along the positive z-axis and therefore has large positive rapidity, whereas the particles in the hadronic system X are moving almost parallel to the negative z-axis (the momentum transfer between the target hadron and the particles in X is small) and they therefore have large negative rapidities.

Events of this kind have been observed by the UA8 collaboration at CERN (Schlein (1993)) and by the H1 (Ahmed *et al.* (1994, 1995a)) and Zeus (Derrick *et al.* (1993, 1995a)) collaborations at DESY. UA8 have measured the energy flow of the particles in the hadronic system X in its rest frame (i.e. the centre-of-mass frame of the hadron–Pomeron system) and observed a substantial peak in the forward direction. This once again suggests that the Pomeron can behave like a point particle, knocking the constituents of the target hadron into the forward direction.

Although the Pomeron seems to behave as though it were a point-like particle, we must remember that it is not a particle at all. It is a Regge trajectory. Nevertheless Ingelman & Schlein (1985) suggested that one can define the structure function of a Pomeron and use diffractive dissociation events to determine the quark and gluon content of the Pomeron. Furthermore, the substructure of the Pomeron has been investigated by the H1 (Ahmed *et al.* (1995b)) and Zeus (Derrick *et al.* (1995b, 1996a)) collaborations.

We shall return to discuss the theory of diffraction dissociation in much more detail in Chapter 7.

1.11 Deep inelastic scattering

The measurement of structure functions ($F_1(x, Q^2)$ and $F_2(x, Q^2)$) in deep inelastic scattering can be thought of as the measurement of the total cross-section for the scattering of an off-shell photon, with square momentum, $-Q^2$, and a proton. The square of the centre-of-mass energy of the photon–proton system is given by

$$s = \frac{Q^2(1 - x)}{x}$$

and so in the Regge limit of $s \gg Q^2$ it follows that $x \ll 1$ (x is the Bjorken-x of the process). At sufficiently low x the off-shellness of the photon is negligible compared with the centre-of-mass energy and so we might expect the total cross-section to have a $1/x$ dependence (at fixed Q^2) similar to the s-dependence of hadronic total cross-sections, i.e. governed by Pomeron exchange. Adding the lower lying meson trajectory, we would then have

$$F_2(x, Q^2) \xrightarrow{x \to 0} A\, x^{-0.08} + B\, x^{0.45}.$$

This fits well for $0.01 \lesssim x \lesssim 0.1$. However, the H1 (Ahmed *et al.* (1995c)) and Zeus (Derrick *et al.* (1995c)) collaborations at HERA have been able to reach values as low as $x \sim 10^{-4}$. The data they obtain show a much steeper x-dependence, e.g. typically $\sim x^{-0.3}$. These data provide, for the first time, evidence of deviations from the Pomeron behaviour described previously.

As we shall see, such deviations are expected within QCD perturbation theory. The large virtuality Q^2 renders a perturbative calculation possible. In Chapter 6 we shall show that perturbative QCD leads to the conclusion that, at sufficiently large Q^2 and sufficiently low x, the structure functions ought to behave like

$$\sim x^{-\omega_0},$$

where

$$\omega_0 = \frac{12\ln 2}{\pi}\alpha_s,$$

and α_s is the strong coupling.

On the other hand, total hadronic cross-sections or low t elastic differential cross-sections cannot be described in terms of perturbative QCD. We expect these processes to be heavily influenced by the non-perturbative properties of QCD, i.e. the Pomeron discussed in this chapter is of non-perturbative origin. We call this the 'soft' Pomeron since in later chapters we shall introduce the concept of the perturbative or 'hard' Pomeron. These are distinct objects. In keeping with modern parlance we use the word 'Pomeron' (soft or hard) in the context of those processes which are characterized by the kinematic condition that the momentum transfer is much smaller than the centre-of-mass energy and in which the vacuum quantum numbers are exchanged.

In future we shall end each chapter with a summary. However, this chapter has been a summary in itself. It has been designed to give the reader sufficient understanding of what we mean when we speak of the Pomeron and why it is an important object. This will be necessary in order to progress through the subsequent chapters, in which we will discuss in detail the question of the reconciliation of Pomeron physics with the 'modern' approach to strong interaction dynamics – namely QCD.

2

A simple example

In this chapter we are going to discuss a simple case in which a quantum field theory simulates the effect of Pomeron exchange in the Regge limit of

$$s \gg |t|.$$

We do not mean that we can identify a Regge trajectory, with associated bound states for various values of positive t, but rather that in this limit the scattering amplitude has the form

$$A(s,t) \propto s^{\alpha_P(t)}. \qquad (2.1)$$

The model we shall consider here is not QCD, but a much simpler quantum field theory, namely a scalar field theory with cubic interactions. We shall show that by summing perturbative contributions to all orders in the coupling constant, but *keeping only leading logarithms*, the behaviour expressed by Eq.(2.1) does indeed emerge. By 'leading logarithms' , we refer to those terms in the perturbative expansion which contain important (in the high energy limit) $\ln s$ factors. Precisely which terms we keep will become clear as we develop the calculation.

An example of Pomeron behaviour from a scalar theory with cubic interactions has been considered before, for example by Polkinghorne (1963a–c) which is described in *The Analytic S-Matrix* by Eden, Landshoff, Olive & Polkinghorne (1966). Their treatment is something more straightforward than the method we shall be introducing here. Feynman diagrams are calculated using the usual method of Feynman parametrization and ladder diagrams are readily summed to all orders. The alternative method that we shall be using here is closer to the treatment by Chang & Yan (1970, 1971). It is something of a sledgehammer to crack a nut. However, the techniques that we shall introduce will serve well in future chapters when they are applied to the more realistic case of QCD.

2.1 The model

We shall represent quarks (and antiquarks) by a complex scalar field ϕ and gluons by another scalar field χ. In order to avoid the difficulty of infra-red divergences (which will be discussed at length in future chapters) we shall assign a mass m to the gluons (whilst leaving the quarks massless). The gluons can interact with themselves as well as with the quarks. A cubic interaction between scalar fields has dimension of mass. In order to introduce a dimensionless coupling constant g, we shall factor out a mass m from the cubic couplings.

A minor complication occurs when considering the analogue of the colour $SU(N)$ group, which is the gauge group of QCD ($N = 3$, but in what follows we keep the number of colours general so as to expose the colour factors explicitly). The self-interaction term in the Lagrangian of the scalar gluons must be symmetric under interchange of two (bosonic) gluons, but we would like the interaction vertex to be proportional to the structure constants of the colour group (which are antisymmetric under interchange of colour indices). This leads us to introduce a colour group which is a product of two $SU(N)$ groups. Thus the gluon fields carry two colour indices and are denoted by $\chi^{a,r}$ with $a, r = 1 \cdots (N^2 - 1)$. The quark field transforms in the fundamental representation of both of these $SU(N)$ groups and so also carries two indices, i.e. $\phi_{i,l}$ with $i, l = 1 \cdots N$. This is rather cumbersome, but in fact the colour factors are in general quite easy to keep track of (and at least there will be some feature which is simpler in QCD!).

Thus the Lagrangian density for this model may be written

$$
\begin{aligned}
\mathcal{L} &= \partial^\mu \bar{\phi}^{i,l} \partial_\mu \phi_{i,l} + \frac{1}{2} \partial^\mu \chi_{a,r} \partial_\mu \chi^{a,r} - \frac{m^2}{2} \chi_{a,r} \chi^{a,r} \\
&\quad - gm \bar{\phi}^{i,l} (T^a)_i^j (T^r)_l^m \phi_{j,m} \chi_{a,r} - \frac{gm}{3!} f_{abc} f_{rst} \chi^{a,r} \chi^{b,s} \chi^{c,t},
\end{aligned}
$$

where the matrices T^a and T^r are the generators of the two $SU(N)$ groups whose structure constants are f_{abc} and f_{rst} respectively.

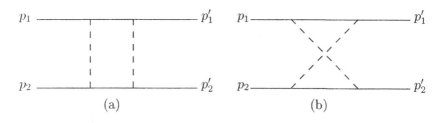

Fig. 2.1. Leading order contribution to Pomeron exchange.

Thus

$$\left[T^a, T^b\right] = i f_{abc} T^c, \quad [T^r, T^s] = i f_{rst} T^t. \qquad (2.2)$$

We do not have an analogue of the quartic coupling between gluons. It turns out that in QCD these interactions always give contributions which are sub-leading in $\ln s$ and we therefore neglect them. We can also assume that the quark fields carry a flavour index which we have suppressed.

Within the context of this model we shall now calculate to all orders in perturbation theory, but keeping the leading powers of $\ln s$ in each order, the process of quark–quark scattering via the exchange of a colour singlet. We assume that the two quarks have different flavours and they emerge from the scattering with the same colour with which they entered.

2.2 The leading order contribution

The leading order Feynman diagrams contributing to this process are shown in Fig. 2.1. The quark lines are denoted by solid lines and the gluons by dashed lines. Because the quarks have different flavours we do not have to consider diagrams with quarks exchanged in the t-channel.

The ingoing quarks have momenta p_1 and p_2, respectively, and the outgoing quarks have momenta p'_1 and p'_2 respectively. Since we are interested in purely elastic scattering we need to consider graphs which do not alter the colour of the incoming quarks, i.e. colour singlet exchange. Therefore there is no contribution to the process in which only one gluon is exchanged and the minimum number of exchanged gluons must be two. The second diagram

(Fig. 2.1(b)) is related to the first by interchange of the incoming and outgoing lower quark lines. The colour generators on the lower line are reversed, but since we are concerned with colour singlet exchange, the two diagrams have the same colour factor. Thus the only difference comes from the kinematics. In other words by crossing symmetry it is sufficient to calculate the contribution from Fig. 2.1(a) and obtain the other contribution from the interchange of the Mandelstam variables s and u (which is equivalent to the interchange of p_2 and p_2').

We deal first with the colour factor. This is straightforward. For a colour singlet exchange we obtain a factor for *each* of the $SU(N)$ groups of

$$\frac{1}{N^2} \text{Tr}(T^a T^b) \text{Tr}(T^a T^b) = \frac{N^2 - 1}{4N^2},$$

giving an overall colour factor of

$$\frac{(N^2 - 1)^2}{16N^4}. \tag{2.3}$$

Fig. 2.1(a) is a one loop diagram, which can be calculated by the conventional means of Feynman parametrization, and the leading logarithm term $\ln(s/t)$ can be extracted from the integral over Feynman parameters. However, it turns out in general to be much more convenient to use dispersive techniques, i.e. we apply the Cutkosky rules (Cutkosky (1960)), which tell us that the imaginary part of this amplitude can be related to a phase-space integral of a product of two amplitudes at the tree level (see Eq.(1.1) and Fig. 1.1), i.e.

$$\Im m \mathcal{A}_{(2.1a)} = \frac{1}{2} \int d\left(P.S.^2\right) \mathcal{A}_0^{(g)}(k) \mathcal{A}_0^{(g)\dagger}(k - q), \tag{2.4}$$

where $\mathcal{A}_0^{(g)}$ is the tree amplitude for single gluon exchange shown either side of the cut in Fig. 2.2, i.e.

$$\mathcal{A}_0^{(g)}(k) = -g^2 m^2 \frac{1}{(k^2 - m^2)}$$

up to a colour factor. $\mathcal{A}_0^{(g)\dagger}$ is the hermitian conjugate of the amplitude, i.e. the complex conjugate of the amplitude with the signs of the momenta reversed. The vector q^μ is the momentum transferred and so $t = q^2$. The symbol $d\left(P.S.^2\right)$ means the integral over

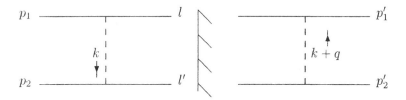

Fig. 2.2. Imaginary part of Fig. 2.1(a). We adopt the convention that t-channel momenta on the left of the cut are directed *downwards*, whereas t-channel momenta on the right of the cut are directed *upwards*.

the phase space of the two cut lines (whose momenta are l and l'), i.e.

$$\int d\left(P.S.^2\right) = \int \frac{d^4l}{(2\pi)^3} \frac{d^4l'}{(2\pi)^3} \delta(l^2)\delta(l'^2)(2\pi)^4\delta^4(p_1 + p_2 - l - l').$$

One of these integrals (say d^4l') can be used to absorb the energy-momentum conserving delta function $\delta^4(p_1 + p_2 - l - l')$ and, for the other, it is convenient to integrate not over the momentum of the other outgoing particle, but over the momentum k of the exchanged gluon. Thus we have

$$\int d\left(P.S.^2\right) = \frac{1}{(2\pi)^2} \int d^4k \; \delta((p_1 - k)^2) \; \delta((p_2 + k)^2).$$

Now we parametrize the momentum k in terms of **Sudakov parameters** ρ and λ:

$$k^\mu = \rho p_1^\mu + \lambda p_2^\mu + k_\perp^\mu,$$

where $k_\perp^\mu$ is the momentum transverse to p_1 and p_2 and we represent this two-dimensional vector by the boldface $\mathbf{k}$. In other words in the centre-of-mass frame in which the incoming particles are considered to be along the z-axis we have

$$p_1^\mu = \left(\frac{\sqrt{s}}{2}, \frac{\sqrt{s}}{2}, \mathbf{0}\right),$$

$$p_2^\mu = \left(\frac{\sqrt{s}}{2}, -\frac{\sqrt{s}}{2}, \mathbf{0}\right),$$

$$k^\mu = \left((\rho + \lambda)\frac{\sqrt{s}}{2}, (\rho - \lambda)\frac{\sqrt{s}}{2}, \mathbf{k}\right).$$

Using $s = 2p_1 \cdot p_2$ and performing the change of variables the phase-space integral becomes

$$\int d\left(P.S.^2\right) = \frac{s}{8\pi^2} \int d\rho \, d\lambda \, d^2\mathbf{k} \, \delta(-s(1-\rho)\lambda - \mathbf{k}^2) \, \delta(s(1+\lambda)\rho - \mathbf{k}^2). \tag{2.5}$$

In the limit $|t| \ll s$ the momentum transferred q^μ is dominated by its transverse component (i.e. $t = q^2 \approx -\mathbf{q}^2$), as can easily be checked from the requirement that the outgoing particles on the right hand side of Fig. 2.2 must be on their mass-shell. Similarly the magnitude of $\mathbf{k}$ will also be of the order of the larger of m and $\sqrt{|t|}$ (it is unlikely that the momentum transferred in the two parts of the diagram on either side of the cut will be much larger than $\sqrt{|t|}$ in such a way that the sum of the two transverse momentum vectors gives $\mathbf{q}$). Thus the delta functions in Eq.(2.5) which give $\lambda = -\rho$ and $\rho \approx \mathbf{k}^2/s$ tell us that both ρ and $|\lambda|$ are both of order $-t/s$ and very much smaller than 1. This means that k^2 may be approximated by

$$k^2 \approx -\mathbf{k}^2$$

and similarly

$$(k - q)^2 \approx -(\mathbf{k} - \mathbf{q})^2.$$

Absorbing the delta functions to perform the integration over ρ and λ:

$$\Im m \mathcal{A}_{(2.1a)} = \frac{(N^2-1)^2}{16N^4} \frac{g^4 m^4}{16\pi^2 s} \int d^2\mathbf{k} \frac{1}{(\mathbf{k}^2 + m^2)} \frac{1}{\left((\mathbf{k} - \mathbf{q})^2 + m^2\right)}. \tag{2.6}$$

The integral over the transverse momentum, $\mathbf{k}$, is readily performed. We choose not to do it here, rather we want to write Eq.(2.6) as

$$\Im m \mathcal{A}_{(2.1a)} = \frac{(N^2-1)^2}{16N^4} \frac{g^4 m^4}{16\pi^2 s} \int d^2\mathbf{k} \, f_0(\mathbf{k}, \mathbf{q}), \tag{2.7}$$

where

$$f_0(\mathbf{k}, \mathbf{q}) = \frac{1}{(\mathbf{k}^2 + m^2)} \frac{1}{\left((\mathbf{k} - \mathbf{q})^2 + m^2\right)}. \tag{2.8}$$

The reason for this apparently perverse notation will become clear when we go on to consider higher order contributions.

The imaginary part then immediately gives us the coefficient of the term $\ln(s/t)$, simply by using the relation (noting that s/t is *negative*):

$$\ln\left(\frac{s}{t}\right) = \ln\left(\frac{s}{|t|}\right) - i\pi.$$

In Eq.(2.6) we have computed the coefficient of the $i\pi$ and the mere existence of an imaginary part tells us that there must be a logarithm in the real part with equal and opposite coefficient. However, we note that when the contribution from Fig. 2.1(b) is added, the large logarithm cancels and we are left with only the imaginary part. This is seen by observing that the contribution to Fig. 2.1(a) is proportional to

$$\frac{1}{s}\left(\ln\left(\frac{s}{|t|}\right) - i\pi\right)$$

and to obtain the contribution from Fig. 2.1(b) we simply replace s by u. Now since u/t is positive this diagram does *not* possess an imaginary part in leading order. It simply has the contribution proportional to

$$\frac{1}{u}\ln\left(\frac{u}{t}\right).$$

Since $u \approx -s$ the logarithms cancel and we are left with the purely imaginary part from Fig. 2.1(a).

2.3 Next-to-leading order contribution

In this section we are interested in those contributions which are of order $g^2 \ln s$ relative to the leading order contribution (calculated in the last section). This means that the vast majority of the higher order graphs can be neglected. The only diagram contributing to the leading logarithm in this order is shown in Fig. 2.3: the so-called one-rung ladder diagram (this will unfortunately *not* be true in the case of QCD). We shall explain why other types of diagram are suppressed at the end of this section. We start as before by considering the colour factor. This gives us a factor of N for each $SU(N)$, relative to the leading order contribution, as can be seen from the relation

$$\text{Tr}(T^c T^d) f_{ace} f_{bde} = N\text{Tr}(T^a T^b). \tag{2.9}$$

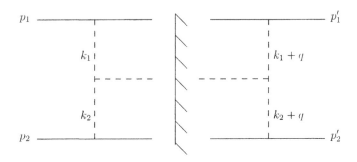

Fig. 2.3. One-rung ladder diagram.

We calculate the imaginary part of this diagram using the same dispersive technique used in the preceding section, i.e.

$$\Im m \mathcal{A}_{(2.3)} = \frac{1}{2} \int d\left(P.S.^3\right) \mathcal{A}_1^{(g)}(k)\mathcal{A}_1^{(g)\dagger}(k-q) \qquad (2.10)$$

where

$$\mathcal{A}_1^{(g)}(k) = g^3 m^3 \frac{1}{(k_1^2 - m^2)(k_2^2 - m^2)}. \qquad (2.11)$$

Once again we write the momenta of the exchanged gluons (k_1 and k_2) in terms of Sudakov variables $\rho_1, \lambda_1, \mathbf{k_1}, \rho_2, \lambda_2, \mathbf{k_2}$, and the three-body phase-space integral becomes

$$\int d\left(P.S.^3\right) = \frac{s^2}{128\pi^5} \int d\rho_1 d\lambda_1 d^2\mathbf{k_1} d\rho_2 d\lambda_2 d^2\mathbf{k_2}$$
$$\delta(-s(1-\rho_1)\lambda_1 - \mathbf{k_1^2})\, \delta(s(1+\lambda_2)\rho_2 - \mathbf{k_2^2})$$
$$\delta(s(\rho_1 - \rho_2)(\lambda_1 - \lambda_2) - (\mathbf{k_1} - \mathbf{k_2})^2). \qquad (2.12)$$

Since $p_1^2 = 0$ and $p_2^2 = 0$, we expect a symmetry in $\mathbf{k_1}$ and $\mathbf{k_2}$, so, as before, we expect all the transverse momenta to have magnitudes which are of the order of the larger of m and $\sqrt{|t|}$. The three-body phase-space integral gives a leading logarithm term $\propto \ln s$ with s scaled by the squared transverse momenta. To leading logarithm order it does not matter exactly what values these transverse momenta that scale the logarithms are. Thus when considering the kinematic limits for the variables $\rho_{1,2}$ and $\lambda_{1,2}$ we can set

$$\mathbf{k_1^2} = \mathbf{k_2^2} = (\mathbf{k_1} - \mathbf{k_2})^2 = \mathbf{k}^2,$$

where $\mathbf{k}$ is a generic transverse momentum whose magnitude is much smaller than $\sqrt{s}$. This situation is in marked contrast to that of deep inelastic scattering (away from very low Bjorken-x), where at one end of the ladder[†] there is a very off-shell photon with squared momentum $-Q^2$ and the leading $\ln Q^2$ contribution is dominated by the region of phase space in which the transverse momenta are strongly ordered up the ladder.

The energies of the cut lines in Fig. 2.3 must be positive *in any Lorentz frame*. This means that the components in the direction of p_1 and p_2 must both be positive for all external lines. This leads to kinematic limits

$$1 > \quad \rho_1 \quad > \rho_2 > 0$$
$$1 > \quad |\lambda_2| \quad > |\lambda_1| > 0$$

$$(2.13)$$

(note that $\lambda_{1,2}$ are negative). We shall argue below that for the leading logarithm these inequalities may be replaced by strong orderings, i.e.

$$1 \gg \quad \rho_1 \quad \gg \rho_2$$
$$1 \gg \quad |\lambda_2| \quad \gg |\lambda_1|.$$

$$(2.14)$$

In this approximation, the three-body phase-space integral may be replaced by

$$\int d\left(P.S.^3\right) \quad = \quad \frac{s^2}{128\pi^5} \int d\rho_1 d\lambda_1 d^2\mathbf{k_1} d\rho_2 d\lambda_2 d^2\mathbf{k_2}$$
$$\times \; \delta(-s\lambda_1 - \mathbf{k}^2)\, \delta(s\rho_2 - \mathbf{k}^2)$$
$$\times \; \delta(-s(\rho_1\lambda_2) - \mathbf{k}^2).$$

$$(2.15)$$

Now performing the integrations over $\lambda_{1,2}$ by absorbing two of the delta functions we end up with

$$\int d\left(P.S.^3\right) = \frac{1}{128\pi^5} \int_{\rho_2}^1 \frac{d\rho_1}{\rho_1} d\rho_2 d^2\mathbf{k_1} d^2\mathbf{k_2} \delta(s\rho_2 - \mathbf{k}^2). \quad (2.16)$$

We can easily perform the integration over ρ_2 by absorbing the remaining delta function and then the $\ln s$ term arises from the integral over ρ_1, i.e.

$$\int_{\mathbf{k}^2/s}^1 \frac{d\rho_1}{\rho_1}.$$

[†] Scaling violations in deep inelastic scattering are driven by ladder diagrams in QCD as embodied in the DGLAP equations (see Chapter 6).

It is this integral which, in the leading logarithm approximation, is dominated by the region $1 \gg \rho_1 \gg \mathbf{k}^2/s$. We can see this by introducing two parameters, ϵ_1 and ϵ_2, such that $1 \gg \epsilon_1, \epsilon_2 \gg \mathbf{k}^2/s$ and splitting the integral up into three parts:

$$
\begin{aligned}
\int_{\mathbf{k}^2/s}^1 \frac{d\rho_1}{\rho_1} &= \left[\int_{\mathbf{k}^2/s}^{\mathbf{k}^2/s\epsilon_1} + \int_{\mathbf{k}^2/s\epsilon_1}^{1/\epsilon_2} + \int_{1/\epsilon_2}^1 \right] \frac{d\rho_1}{\rho_1} \\
&= -\ln \epsilon_1 + \left(\ln\left(\epsilon_1/\epsilon_2\right) + \ln\left(s/\mathbf{k}^2\right)\right) + \ln \epsilon_2 .
\end{aligned}
$$

Since $s/\mathbf{k}^2 \gg 1/\epsilon_1, 1/\epsilon_2$ this is dominated by the middle part of the integral for which $1 \gg \rho_1 \gg \mathbf{k}^2/s$, as required. This argument may seem a little far fetched, since we are assuming that the ϵ_i are sufficiently large compared with $\mathbf{k}^2/s$ that we can neglect their logarithms, and it might be felt that this only works when s is extremely large. Nevertheless this is the formal definition of the leading logarithm approximation and corrections are indeed suppressed by powers of $\ln s$. Thus we have justified the assumption of strong ordering in the ρ s which, together with the on-shell conditions for the cut lines, give a similar strong ordering (in the opposite direction) for the λ s, thereby justifying the strong inequality Eq.(2.14).

Since we now have $s\rho_1\lambda_2 \approx \mathbf{k}^2$, it follows that

$$
\begin{aligned}
s\rho_1\lambda_1 &\ll \mathbf{k}_1^2 \\
s\rho_2\lambda_2 &\ll \mathbf{k}_2^2
\end{aligned}
$$

so that $\mathcal{A}_1$ (Eq.(2.11)) may be rewritten

$$
\mathcal{A}_1^{(g)}(k) = g^3 m^3 \frac{1}{(\mathbf{k}_1^2 + m^2)(\mathbf{k}_2^2 + m^2)} . \qquad (2.17)
$$

Now we introduce f_1 in analogy with f_0 (Eq.(2.7)), i.e.

$$
\Im m \mathcal{A}_{(2.3)} = \frac{(N^2 - 1)}{16 N^4} \frac{g^4 m^4}{16 \pi^2 s} \int d^2\mathbf{k}_1 f_1(s, \mathbf{k}_1, \mathbf{q}), \qquad (2.18)
$$

where

$$
f_1(s, \mathbf{k}_1, \mathbf{q}) = \frac{g^2 m^2 N^2 s}{2(2\pi)^3} \int_0^1 d\rho_2 \int_{\rho_2}^1 \frac{d\rho_1}{\rho_1} \delta(s\rho_2 - \mathbf{k}^2) \int d^2\mathbf{k}_2
$$

$$
\times \frac{1}{(\mathbf{k}_1^2 + m^2)(\mathbf{k}_2^2 + m^2)} \frac{1}{((\mathbf{k}_1 - \mathbf{q})^2 + m^2)((\mathbf{k}_2 - \mathbf{q})^2 + m^2)} . \qquad (2.19)
$$

With a view to application in the more complicated case of QCD, rather than performing the ρ-integral, we introduce the technique of Mellin transforms (a survival kit on Mellin transforms appears in the appendix to this chapter). This has the effect of unravelling the nested integrals in the ρs. Thus we define the Mellin transform of $f_1(s, \mathbf{k_1}, \mathbf{q})$ to be $\mathcal{F}_1(\omega, \mathbf{k_1}, \mathbf{q})$ given by

$$\mathcal{F}_1(\omega, \mathbf{k_1}, \mathbf{q}) = \int_1^\infty d\left(\frac{s}{\mathbf{k}^2}\right) \left(\frac{s}{\mathbf{k}^2}\right)^{-\omega-1} f_1(s, \mathbf{k_1}, \mathbf{q}).$$

In this definition we have normalized s by the square of the typical transverse momentum, $\mathbf{k}$, in order to be able to keep track of dimensions. Recall that for the leading logarithm approximation the exact normalization does not matter as long as it is a scale which is small compared with s.

We perform the integration over s, and obtain

$$\mathcal{F}_1(\omega, \mathbf{k_1}, \mathbf{q}) = \frac{g^2 m^2 N^2}{2(2\pi)^3} \int_0^1 d\rho_2 \int_{\rho_2}^1 \frac{d\rho_1}{\rho_1} \rho_2^{\omega-1} \int d^2 \mathbf{k_2}$$

$$\times \frac{1}{(\mathbf{k_1}^2 + m^2)(\mathbf{k_2}^2 + m^2)} \frac{1}{((\mathbf{k_1} - \mathbf{q})^2 + m^2)((\mathbf{k_2} - \mathbf{q})^2 + m^2)}. \quad (2.20)$$

The integrations over the ρs are unravelled by the change of variables

$$\tau_1 = \rho_1$$
$$\tau_1 \tau_2 = \rho_2.$$

The limits of integration are now simply

$$0 < \tau_{1,2} < 1$$

and the Jacobian for this change of integration variables is ρ_1, so we obtain

$$\mathcal{F}_1(\omega, \mathbf{k_1}, \mathbf{q}) = \frac{g^2 m^2 N^2}{2(2\pi)^3} \int_0^1 d\tau_1 \, \tau_1^{\omega-1} \int_0^1 d\tau_2 \, \tau_2^{\omega-1} \int d^2 \mathbf{k_2}$$

$$\times \frac{1}{(\mathbf{k_1}^2 + m^2)(\mathbf{k_2}^2 + m^2)} \frac{1}{((\mathbf{k_1} - \mathbf{q})^2 + m^2)((\mathbf{k_2} - \mathbf{q})^2 + m^2)},$$

i.e.

$$\omega^2 \mathcal{F}_1(\omega, \mathbf{k_1}, \mathbf{q}) = \frac{g^2 m^2 N^2}{2(2\pi)^3} \int d^2 \mathbf{k_2}$$

$$\times \frac{1}{(\mathbf{k_1}^2 + m^2)(\mathbf{k_2}^2 + m^2)} \frac{1}{((\mathbf{k_1} - \mathbf{q})^2 + m^2)((\mathbf{k_2} - \mathbf{q})^2 + m^2)}. \quad (2.21)$$

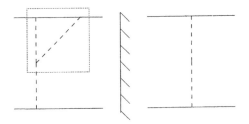

Fig. 2.4. A vertex correction diagram.

We shall write this in a suggestive form as

$$\omega \mathcal{F}_1(\omega, \mathbf{k_1}, \mathbf{q}) \;=\; \frac{g^2 m^2 N^2}{2(2\pi)^3} \int d^2 \mathbf{k_2}$$

$$\times \; \frac{1}{(\mathbf{k_2^2} + m^2)} \frac{1}{\left((\mathbf{k_2} - \mathbf{q})^2 + m^2\right)} \mathcal{F}_0(\omega, \mathbf{k_1}, \mathbf{q}), \quad (2.22)$$

where $\mathcal{F}_0(\omega, \mathbf{k_1}, \mathbf{q}) \;=\; \omega^{-1} f_0(\mathbf{k_1}, \mathbf{q})$ is the Mellin transform of $f_0(\mathbf{k_1}, \mathbf{q})$, given in Eq.(2.8).

An example of a diagram that has been neglected is shown in Fig. 2.4, which is a vertex correction to the leading order contribution. This certainly contains an extra g^2 relative to the leading order graph, but no extra $\ln s$, since the vertex correction (shown in the dotted box in Fig. 2.4) cannot depend upon s as the squared momentum of the lines coming into the vertex is either zero or k^2, which is of order t (i.e. the on-shell condition of the cut upper quark line means we cannot strongly order the Sudakov components of the t-channel gluons). This is the case for all diagrams which have vertex or self-energy insertions.

There are also other diagrams which one can draw to this order which do not contribute in the leading logarithm approximation. The first is shown in Fig. 2.5, which is a vertex correction diagram, but with three cut lines. The momenta k_1 and k_2 are still ordered as discussed above, so

$$|\lambda_2| \gg \frac{\mathbf{k}^2}{s}$$

and the squared momentum of the upper quark line on the right hand side of the cut, $(p_1 - k_2)^2$, is of order $|\lambda_2|s \gg \mathbf{k}^2$. This highly virtual quark will give a large denominator (compared with the

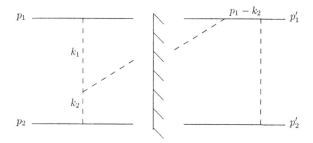

Fig. 2.5. A (cut) vertex correction diagram.

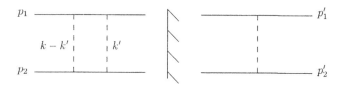

Fig. 2.6. A three gluon exchange diagram.

denominators from Fig. 2.3, which are all of order $\mathbf{k}^2$) and the graph is therefore suppressed and does not contribute in leading logarithm approximation. This is a feature of the scalar theory and does not hold in the case of QCD, where momenta arising from the vertices can compensate for this hard propagator. Furthermore, we neglect diagrams in which there are fermion loops (e.g. a diagram in which there are three quarks and an antiquark rather than two quarks and two gluons in the intermediate state). In the present case we argue that the colour factor is suppressed by $1/N^2$. However, in the case of QCD we shall argue in the next chapter that all such fermion loop diagrams are sub-leading in $\ln s$.

The other type of diagram that we have to consider is the three gluon exchange diagram, which is shown in Fig. 2.6. In the diagram the cut is to the right of two of the gluons (there is also a contribution in which the two gluons on the left of the cut are crossed, and a further contribution to the imaginary part of the diagram where the cut is to the left of two gluons). However, for this type of diagram there is very little phase space when all the

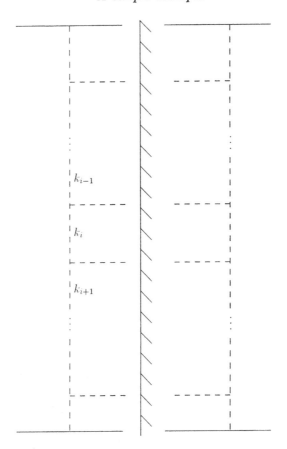

Fig. 2.7. n-rung ladder diagram.

denominators are small and so the amplitudes are suppressed by a power of m^2/s, compared with diagrams with only two gluons exchanged. These diagrams may therefore be neglected. This is also a feature which holds in the scalar theory but not in QCD.

2.4 The n-rung ladder diagram

It is now relatively straightforward to generalize the above discussion to any order in perturbation theory. The order $(g^2 \ln s)^n$ correction to the leading order approximation is given by the n-rung uncrossed ladder diagram (Fig. 2.7) whose amplitude has an imaginary part:

$$\Im m \mathcal{A}_{(2.7)} = \frac{1}{2} \int d\left(P.S.^{(n+2)}\right) \mathcal{A}_n^{(g)}(k) \mathcal{A}_n^{(g)\dagger}(k-q) \qquad (2.23)$$

where

$$\mathcal{A}_n^{(g)}(k) = (-gm)^{n+2} \prod_{i=1}^{n+1} \frac{1}{(k_i^2 - m^2)} \qquad (2.24)$$

up to a colour factor. The group theory (see Eq.(2.9)) gives a factor of N^2 for each rung relative to the leading order colour factor (Eq.(2.3)).

The momentum of the ith upright section of the ladder is written

$$k_i^\mu = \rho_i p_1^\mu + \lambda_i p_2^\mu + k_{i\perp}^\mu$$

with

$$k_{i\perp}^\mu = (0, 0, \mathbf{k_i})$$

and the $(n+2)$-body phase-space integral is then

$$\int d\left(P.S.^{(n+2)}\right) = \frac{s^{n+1}}{2^{4n+3}\pi^{3n+2}} \int \prod_{i=1}^{n+1} d\rho_i d\lambda_i d^2 \mathbf{k_i}$$

$$\times \prod_{j=1}^{n} \delta\big(s(\rho_j - \rho_{j+1})(\lambda_j - \lambda_{j+1}) - (\mathbf{k_j} - \mathbf{k_{j+1}})^2\big)$$

$$\times \delta\big(-s(1-\rho_1)\lambda_1 - \mathbf{k_1^2}\big)\delta\big(s(1+\lambda_{n+1})\rho_{n+1} - \mathbf{k_{n+1}^2}\big). \qquad (2.25)$$

Again the symmetry between p_1 and p_2 (the top and the bottom of the ladder) tells us that the phase-space integral is dominated by the region in which the transverse momenta of the vertical lines (and the horizontal cut lines) are all of order $\mathbf{k}^2$, which is of the order of the larger of m^2 and $|t|$. Furthermore the integral over the Sudakov variables ρ_i and λ_i comes from the region

$$\rho_i \gg \rho_{i+1}$$
$$|\lambda_{i+1}| \gg |\lambda_i|$$

and in this region we have $k_i^2 \approx -\mathbf{k_i^2}$, so that $\mathcal{A}_n^{(g)}(k)$ may be written

$$\mathcal{A}_n^{(g)}(k) = -(gm)^{n+2} \prod_{i=1}^{n+1} \frac{1}{(\mathbf{k_i^2} + m^2)}. \qquad (2.26)$$

The phase-space integral (after integrating the λ_i by absorbing the delta functions which put the cut lines on mass-shell) is

$$
\int d\left(P.S.^{(n+2)}\right) = \frac{1}{2^{4n+3}\pi^{3n+2}} \prod_{i=1}^{n} \int_{\rho_{i+1}}^{1} \frac{d\rho_i}{\rho_i} \prod_{j=1}^{n+1} d^2 \mathbf{k_j}
$$
$$
\times \, d\rho_{n+1} \, \delta(s\rho_{n+1} - \mathbf{k}^2). \qquad (2.27)
$$

The nested integrals over the ρ_i give the leading logarithm contribution proportional to $(\ln s)^n/n!$. We define f_n, in analogy with f_0 and f_1, by

$$
\Im m \mathcal{A}_{(2.7)} = \frac{(N^2-1)^2}{16N^4} \frac{g^4 m^4}{16\pi^2 s} \int d^2 \mathbf{k_1} f_n(s, \mathbf{k_1}, \mathbf{q}) \qquad (2.28)
$$

where

$$
f_n(s, \mathbf{k_1}, \mathbf{q}) = \left(\frac{g^2 m^2 N^2}{2(2\pi)^3}\right)^n \prod_{i=1}^{n} \int_{\rho_{i+1}}^{1} \frac{d\rho_i}{\rho_i} \int_{0}^{1} d\rho_{n+1} \prod_{j=2}^{n+1} d^2 \mathbf{k_j}
$$
$$
\times \prod_{m=1}^{n+1} \frac{1}{(\mathbf{k_m^2} + m^2)((\mathbf{k_m} - \mathbf{q})^2 + m^2)} \, s \, \delta(s\rho_{n+1} - \mathbf{k}^2). \qquad (2.29)
$$

We now take the Mellin transform, integrate over s (absorbing the remaining delta function) and change variables from ρ_i to τ_i, where

$$
\tau_i = \frac{\rho_i}{\rho_{i-1}}
$$

(with $\rho_0 = 1$). The limits on the τ integrals are $0 < \tau_i < 1$ and the Jacobian for this change of integration variables is $\rho_1 \rho_2 \cdots \rho_n$. Hence

$$
\mathcal{F}_n(\omega, \mathbf{k_1}, \mathbf{q}) = \left(\frac{g^2 m^2 N^2}{2(2\pi)^3}\right)^n \prod_{i=1}^{n+1} \int_{0}^{1} \tau_i^{\omega-1} \, d\tau_i \prod_{j=2}^{n+1} d^2 \mathbf{k_j}
$$
$$
\times \prod_{m=1}^{n+1} \frac{1}{(\mathbf{k_m^2} + m^2)((\mathbf{k_m} - \mathbf{q})^2 + m^2)}
$$
$$
= \left(\frac{g^2 m^2 N^2}{2(2\pi)^3}\right)^n \frac{1}{\omega^{n+1}} \left(\int d^2 \mathbf{k} \frac{1}{(\mathbf{k}^2 + m^2)((\mathbf{k} - \mathbf{q})^2 + m^2)}\right)^n
$$
$$
\times \frac{1}{(\mathbf{k_1^2} + m^2)((\mathbf{k_1} - \mathbf{q})^2 + m^2)}. \qquad (2.30)
$$

Note that the factor $(1/\omega)^{n+1}$ is the Mellin transform of $(\ln s)^n/n!$.

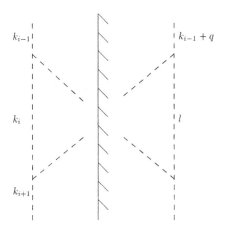

Fig. 2.8. A section of a crossed ladder diagram.

Crossed ladder diagrams do not contribute in leading logarithm approximation. A section of such a ladder is shown in Fig. 2.8. In this diagram the momentum on the right of the cut, marked l, is given by

$$l = k_{i-1} + k_{i+1} - k_i - q.$$

In the limit $\rho_{i-1} \gg \rho_i \gg \rho_{i+1}$ and $|\lambda_{i+1}| \gg |\lambda_i| \gg |\lambda_{i-1}|$, this propagator gives rise to a denominator which is of order

$$l^2 \approx s\lambda_{i+1}\rho_{i-1}$$

but $s\lambda_{i+1}$ is of order $\mathbf{k}^2/\rho_i$ (from the mass-shell condition of the ith cut line) and so we have

$$l^2 \approx \frac{\rho_{i-1}}{\rho_i}\mathbf{k}^2,$$

which is much larger than $\mathbf{k}^2$ (since $\rho_{i-1} \gg \rho_i$). Thus there is a large denominator, which suppresses the contribution from this diagram so that it no longer contributes in leading logarithm approximation. Once again QCD does not possess this rather convenient feature.

The series, $\sum_{n=0}^{\infty} \mathcal{F}_n(\omega, \mathbf{k_1}, \mathbf{q})$, is a simple geometrical series (see Eq.(2.30)) and can be summed to give

$$\mathcal{F}(\omega, \mathbf{k}, \mathbf{q}) = \frac{1}{(\mathbf{k}^2 + m^2)((\mathbf{k} - \mathbf{q})^2 + m^2)(\omega - 1 - \alpha_P(t))}, \quad (2.31)$$

where

$$\alpha_P(t) = -1 + \frac{g^2 m^2 N^2}{2(2\pi)^3} \int d^2 \mathbf{k}' \frac{1}{(\mathbf{k}'^2 + m^2)((\mathbf{k}' - \mathbf{q})^2 + m^2)} \qquad (2.32)$$

(with $t = -\mathbf{q}^2$). The integral over the transverse momentum is readily computed. For small t ($|t| \ll m^2$) we have

$$\alpha_P(t) \approx -1 + \frac{g^2 N^2}{16\pi^2}\left(1 + \frac{t}{6m^2}\right). \qquad (2.33)$$

The trajectory rapidly becomes non-linear as $|t|$ becomes of order m^2. Thus we see that $\mathcal{F}(\omega, \mathbf{k}, \mathbf{q})$ has a simple pole in ω at $\omega = 1 + \alpha_P(t)$.

2.5 The integral equation

Although we already have a solution for $\mathcal{F}(\omega, \mathbf{k}, \mathbf{q})$, in preparation for the case of QCD it is useful to establish an integral equation which gives the same result. Such an integral equation is shown schematically in Fig. 2.9. It is an implicit equation with $\mathcal{F}(\omega, \mathbf{k}, \mathbf{q})$ appearing on both sides. Basically it tells us that $\mathcal{F}$ is equal to the leading order term plus $\mathcal{F}$ with an extra rung added. The extra rung introduces a coupling constant factor of $g^2 m^2$, a colour factor of N^2, two propagators for the extra internal lines, $1/(\mathbf{k}'^2 + m^2)$ and $1/((\mathbf{k}' - \mathbf{q})^2 + m^2)$, and an extra phase-space integral, which in the Mellin transform representation gives a factor of $1/(2(2\pi)^3\omega)$ combined with an integral over the transverse momentum $d^2\mathbf{k}'$. Thus the integral equation is

$$\omega \mathcal{F}(\omega, \mathbf{k}, \mathbf{q}) = \frac{1}{(\mathbf{k}^2 + m^2)((\mathbf{k} - \mathbf{q})^2 + m^2)}$$
$$+ \frac{g^2 m^2 N^2}{2(2\pi)^3\omega} \int d^2 \mathbf{k}' \frac{\mathcal{F}(\omega, \mathbf{k}, \mathbf{q})}{(\mathbf{k}'^2 + m^2)((\mathbf{k}' - \mathbf{q})^2 + m^2)}. \qquad (2.34)$$

We see that if we insert the first term on the right hand side into $\mathcal{F}$ in the second term, we obtain the one-rung ladder contribution, and inserting this into $\mathcal{F}$ in the second term gives the two-rung contribution, etc. By iteration we thus see that the integral equation generates all the ladder diagrams.

The integral equation of course gives the same solution as Eq.(2.31).

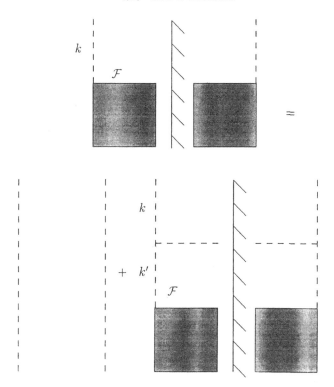

Fig. 2.9. The integral equation.

2.6 The Pomeron

After this rather tortuous route we now come to the solution for the amplitude $\mathcal{A}(s,t)$ for the colour singlet exchange. Inverting the Mellin transform we have

$$\Im m\mathcal{A}(s,t) = \frac{(N^2-1)^2}{16N^4}\frac{g^4m^4}{16\pi^2s}$$
$$\times \int d^2\mathbf{k}\frac{1}{(\mathbf{k}^2+m^2)((\mathbf{k}-\mathbf{q})^2+m^2)}\left(\frac{s}{|t|}\right)^{1+\alpha_P(t)}, (2.35)$$

with $\alpha_P(t)$ given by Eq.(2.32) and we have substituted $|t|$ for $\mathbf{k}^2$ in the normalization of s, which we may do without affecting the leading logarithms.

Let us write this as

$$\Im m \mathcal{A}(s,t) = \frac{C}{s}\left(\frac{s}{|t|}\right)^{\epsilon_P(t)},$$

where

$$\epsilon_P(t) = 1 + \alpha_P(t)$$

(note that $\epsilon_P(t)$ is of order g^2). Up to corrections which are of order g^2, this is the imaginary part of

$$
\begin{aligned}
\mathcal{A}(s,t) &= \left(\frac{s}{-t}\right)^{\epsilon_P(t)} \frac{C}{s\pi\epsilon_P(t)}\left[\cos\pi\epsilon_P(t) + i\sin\pi\epsilon_P(t)\right] \\
&\approx \frac{C}{\pi\epsilon_P(t)s}\left(\frac{s}{t}\right)^{\epsilon_P(t)}.
\end{aligned}
$$

Remember that we must add the contribution from the crossed amplitude in which s is replaced by u. Thus the entire contribution is

$$
\begin{aligned}
\mathcal{A}(s,t) &= \frac{C}{\pi\epsilon_P(t)s}\left(\frac{s}{t}\right)^{\epsilon_P(t)} \\
&+ \frac{C}{\pi\epsilon_P(t)u}\left(\frac{u}{t}\right)^{\epsilon_P(t)}.
\end{aligned}
$$

In the Regge limit $u \approx -s$ and so we see that the real parts cancel in leading logarithm order and we are left with an amplitude that is *purely imaginary* and given by Eq.(2.35).

We have thus succeeded in deriving the Pomeron in this particular field theory.

2.7 Summary

Let us summarize the important features of Pomeron exchange in the scalar model discussed in this chapter.

• In the scalar model with cubic interactions described in Section 2.1, the leading logarithm contributions to the imaginary part of the amplitude come from uncrossed ladder diagrams, with a cut through the rungs. The cut lines are integrated over the relevant phase space.

• We use Sudakov variables to describe the momentum k_i of the ith vertical line on the left of the ladder by

$$k_i^\mu = \rho_i p_1^\mu + \lambda_i p_2^\mu + k_{i\perp}^\mu$$

with

$$k^\mu_{i\perp} = (0, 0, \mathbf{k_i}).$$

For the right hand side of the ladder the transverse momentum $\mathbf{k_i}$ is replaced by $(\mathbf{k_i} - \mathbf{q})$ (in the Regge limit, $|t| \ll s$).

• The phase-space integral is dominated by the region in which the transverse momenta all have the same order of magnitude, which is denoted by $\mathbf{k}$, such that $\mathbf{k}^2$ is of the order of the larger of m^2 and $|t|$.

• The leading logarithm part of the integral over the longitudinal components comes from the region

$$\rho_i \gg \rho_{i+1}$$
$$|\lambda_{i+1}| \gg |\lambda_i|$$

and in this region the momenta of the vertical lines are dominated by their transverse components so that $k_i^2 \approx -\mathbf{k_i}^2$.

• After integrating over the λ_i and absorbing the delta functions which give the on-shell condition for the cut lines, the remaining integration over the ρ_i are nested integrals which are easily unravelled by taking the Mellin transform.

• An integral equation can be established for the Mellin transform of the imaginary part of the amplitude. The sum of all ladder diagrams is generated if the integral equation is solved iteratively.

• The integral equation has a solution for which the Mellin transform has a simple pole at $\omega = 1 + \alpha_P(t)$, where $\alpha_P(t)$ is given by Eq.(2.32).

• The real part of the amplitude is readily reconstructed from the imaginary part. However when the contribution from the crossed process obtained by interchanging s and u is added, the leading order contribution of the real part cancels, leaving a purely imaginary amplitude.

2.8 Appendix

Mellin transforms

Definition:
The Mellin transform, $\mathcal{F}(\omega)$ of the function $f(s)$ is given by

$$\mathcal{F}(\omega) = \int_1^\infty d\left(\frac{s}{\mathbf{k}^2}\right) \left(\frac{s}{\mathbf{k}^2}\right)^{-\omega-1} f(s) \qquad (A.2.1)$$

and its inverse is given by

$$f(s) = \frac{1}{2\pi i} \int_C d\omega \left(\frac{s}{\mathbf{k}^2}\right)^\omega \mathcal{F}(\omega), \qquad (A.2.2)$$

where the contour C is to the right of all ω-plane singularities of $\mathcal{F}(\omega)$.

Useful examples:
If $f(s)$ is of the form

$$f(s) = s^\alpha g(s),$$

then the Mellin transform $\mathcal{F}(\omega)$ is given by

$$\mathcal{F}(\omega) = \left(\mathbf{k}^2\right)^\alpha \mathcal{G}(\omega - \alpha), \qquad (A.2.3)$$

where $\mathcal{G}(\omega)$ is the Mellin transform of $g(s)$.

If $g(s) = (\ln s)^r$ then its Mellin transform is given by

$$\mathcal{G}(\omega) = \int_1^\infty d\left(\frac{s}{\mathbf{k}^2}\right) \left(\frac{s}{\mathbf{k}^2}\right)^{-\omega-1} (\ln s)^r.$$

Changing variables to $y = \omega \ln(s/\mathbf{k}^2)$ we obtain

$$\mathcal{G}(\omega) = \frac{1}{\omega^{r+1}} \int_0^\infty y^r e^{-y} dy.$$

The integral on the right hand side is the integral definition of the Euler gamma function, $\Gamma(r+1)$ $(= r!$ for integer $r)$. Therefore,

$$\mathcal{G}(\omega) = \frac{\Gamma(r+1)}{\omega^{r+1}}. \qquad (A.2.4)$$

Combining these two results (Eqs.(A.2.3, A.2.4)) we obtain, for the Mellin transform of the function

$$f(s) = (\ln s)^r s^\alpha,$$

$$\mathcal{F}(\omega) = \left(\mathbf{k}^2\right)^\alpha \frac{\Gamma(r+1)}{(\omega - \alpha)^{r+1}}. \qquad (A.2.5)$$

Thus we see that if the function $f(s)$ is a pure power of s, then the Mellin transform has a singularity which is a simple pole. If the function $f(s)$ is a power multiplied by (in general non-integer) powers of $\ln s$, then the Mellin transform has a cut singularity. The factor $\left(\mathbf{k}^2\right)^\alpha$ simply adjusts the dimension. For the high energy behaviour, we are interested in the position and nature of the ω-plane singularities. Note that the Mellin transform of a constant, C, is C/ω.

It is important to be familiar with these relations in *both* directions, i.e. to be able to perform the inverse Mellin transforms and obtain the s-dependence of amplitudes from the singularity structure of the Mellin transforms.

Convolutions:
Let $f(s)$ be given in terms of a convolution of a set of n functions, $f_i(s/\mathbf{k}^2)$ $(i = 1 \cdots n)$, by

$$f(s) = \mathbf{k}^2 \prod_{i=1}^{n} \int_{\rho_{i+1}}^{1} \frac{d\rho_i}{\rho_i} f_i\left(\frac{\rho_{i-1}}{\rho_i}\right) \delta(\rho_n s - \mathbf{k}^2) \qquad (A.2.6)$$

(with $\rho_0 = 1$ and $\rho_{n+1} = 0$). The Mellin transform is given by

$$\mathcal{F}(\omega) = \mathbf{k}^2 \int_1^{\infty} d\left(\frac{s}{\mathbf{k}^2}\right) \left(\frac{s}{\mathbf{k}^2}\right)^{-\omega-1}$$
$$\times \prod_{i=1}^{n} \int_{\rho_{i+1}}^{1} \frac{d\rho_i}{\rho_i} f_i\left(\frac{\rho_{i-1}}{\rho_i}\right) \delta(\rho_n s - \mathbf{k}^2).$$

Performing the integration over $s/\mathbf{k}^2$ (absorbing the delta function) gives

$$\mathcal{F}(\omega) = \prod_{i=1}^{n} \int_{\rho_{i+1}}^{1} \frac{d\rho_i}{\rho_i} f_i\left(\frac{\rho_{i-1}}{\rho_i}\right) \rho_n^{\omega}.$$

Now change variables from ρ_i to τ_i, where

$$\tau_i = \frac{\rho_i}{\rho_{i-1}},$$

so that $\rho_n = \tau_1 \tau_2 \cdots \tau_n$. The Jacobian for the change of variables is $\rho_1 \rho_2 \cdots \rho_{n-1}$, and we finally obtain

$$\mathcal{F}(\omega) = \prod_{i=1}^{n} \int_0^1 d\tau_i \tau_i^{\omega-1} f_i\left(\frac{1}{\tau_i}\right) = \prod_{i=1}^{n} \mathcal{F}_i(\omega), \qquad (A.2.7)$$

where $\mathcal{F}_i(\omega)$ are the Mellin transforms of the functions $f_i(s/\mathbf{k}^2)$.

3

The reggeized gluon

A particle of mass M and spin J is said to 'reggeize' if the amplitude, $\mathcal{A}$, for a process involving the exchange in the t-channel of the quantum numbers of that particle behaves asymptotically in s as

$$\mathcal{A} \propto s^{\alpha(t)}$$

where $\alpha(t)$ is the trajectory and $\alpha(M^2) = J$, so that the particle itself lies on the trajectory.

The idea that particles should reggeize has a long history. It was first proposed by Gell-Mann *et al.* (1962, 1964a,b) and by Polkinghorne (1964). Mandelstam (1965) gave general conditions for reggeization to occur and this was developed by several authors (Abers & Teplitz (1967), Abers *et al.* (1970), Dicus & Teplitz (1971), Grisaru, Schnitzer, & Tsao (1973)). Calculations in Quantum Electrodynamics (QED) were carried out by Frolov, Gribov & Lipatov (1970, 1971) and by Cheng & Wu (1965, 1969a–c, 1970a,b), who showed that the photon had a fixed cut singularity (as opposed to a Regge pole). On the other hand McCoy & Wu (1976a–f) established that the fermion does indeed reggeize in QED. This was extended to non-abelian gauge theories by Mason (1976a,b) and Sen (1983). The demonstration of reggeization of the gluon was first shown to two-loop order by Tyburski (1976), Frankfurt & Sherman (1976), and Lipatov (1976) and to three loops by Cheng & Lo (1976). The reggeization to all orders in perturbation theory has been established by several authors using somewhat different techniques. Mason (1977) worked in Coulomb gauge and used time ordered perturbation theory to establish that the amplitude factorized in such a way that the reggeization must follow. Cheng & Lo (1977) developed a recursion relation for going to higher orders in perturbation theory.

The method that we shall follow in this chapter is that of Fadin,

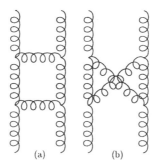

Fig. 3.1. Section of uncrossed and crossed gluon ladder diagrams.

Kuraev & Lipatov (1976), who used dispersive techniques developed in the preceding chapter. We feel that this is the most transparent derivation of reggeization and lends itself most easily to the discussion of the Pomeron in the next chapter.

In the preceding chapter we showed that in a ϕ^3 theory the amplitude for elastic scattering of scalar 'quarks' was dominated in the leading $\ln s$ approximation by uncrossed ladder diagrams. In particular, it was shown that a crossed rung gives rise to a hard denominator and is suppressed by $\sim \rho_i/\rho_{i-1}$. In QCD this does not work. A section of a ladder shown in Fig. 3.1(a) does not dominate over the crossed-rung section shown in Fig. 3.1(b). The reason for this is that the triple gluon vertices carry the momenta of the gluons in the numerators and in Fig. 3.1(b) the scalar product of these momenta between the top left and bottom right (or *vice versa*) vertices produces a term which is enhanced compared with the corresponding scalar product in Fig. 3.1(a). This enhancement compensates for the denominator suppression due to the hard propagator in the crossed-rung diagram.

Nevertheless we shall show that it is possible to organize high energy scattering amplitudes into 'effective' ladder-type diagrams. The vertices will not be the usual triple gluon vertices, but, rather, a non-local effective vertex, which we shall discuss below. Also the vertical lines of the ladder are not bare gluons whose propagators are given (in Feynman gauge) by

$$D_{\mu\nu}(q^2) = -i\frac{g_{\mu\nu}}{q^2}$$

but, rather, they are 'reggeized' gluons whose propagator (in Feynman gauge) is

$$\tilde{D}_{\mu\nu}(\hat{s}, q^2) = -i\frac{g_{\mu\nu}}{q^2}\left(\frac{\hat{s}}{\mathbf{k}^2}\right)^{\epsilon_G(q^2)}, \tag{3.1}$$

where $\sqrt{\hat{s}}$ is the centre-of-mass energy of the particles between which the 'reggeized' gluon is exchanged and $\alpha_G(q^2) = 1 + \epsilon_G(q^2)$ is the Regge trajectory of the gluon.[†]

In order to show that gluons reggeize in this way (and to determine the Regge trajectory) we need to calculate to all orders in the perturbation series but keeping only the leading $\ln s$ terms at each order. We need to select those diagrams in which the exchanged quantum numbers (in the t-channel) are those of the gluon, i.e. spin-1 and colour octet. As discussed in Chapter 1, the amplitude in which a single particle of spin J is exchanged has a large s behaviour proportional to s^J, so we are interested in the contributions to the amplitude which at order α_s^n are proportional to $s\,\alpha_s^n \ln^{n-1}s$ and we shall drop sub-leading logarithm terms. We shall begin by discussing the first three orders of perturbation theory and then generalize to all orders.

3.1 Leading order calculation

The QCD process we consider is the scattering of two quarks with different flavours due to colour octet exchange and within the Regge limit ($s \gg -t$). We neglect the masses of the quarks and assume that their incoming momenta p_1 and p_2 lie along the z-axis, i.e.

$$p_1 = \frac{\sqrt{s}}{2}(1, 1, \mathbf{0}),$$

$$p_2 = \frac{\sqrt{s}}{2}(1, -1, \mathbf{0}).$$

The tree diagram contribution to this amplitude is shown in Fig. 3.2.[‡] It is very important to realize that all the components of the momentum of the exchanged gluon, q^μ, are much smaller than $\sqrt{s}$. This is true because we are interested in the region $|q^2| = |t| \ll s$

[†] As in the preceding chapter $\mathbf{k}^2$ represents a typical transverse momentum.
[‡] We use the Feynman rules for QCD given in the appendix at the end of the book.

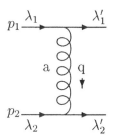

Fig. 3.2. Tree level amplitude.

and because the outgoing quarks are on mass-shell (i.e. $(p_1 - q)^2 = 0$ and $(p_2 + q)^2 = 0$).

3.1.1 The eikonal approximation

The eikonal approximation is an extremely important ingredient in building the 'reggeized' gluon and subsequently the QCD Pomeron.

The upper line of the diagram in Fig. 3.2 gives the factor

$$-ig\,\bar{u}(\lambda_1', p_1 - q)\gamma^\mu u(\lambda_1, p_1)\tau^a_{ij}$$

(where λ_1, λ_1' are the helicities of the incoming and outgoing quarks respectively and the τ^a are the generators of the colour group in the fundamental representation). Since all the components of q^μ are small we may replace this by

$$-ig\,\bar{u}(\lambda_1', p_1)\gamma^\mu u(\lambda_1, p_1)\tau^a_{ij}.$$

For spinors normalized such that $u^\dagger(\lambda_1', p_1)u(\lambda_1, p_1) = 2E_{p_1}\delta_{\lambda_1'\lambda_1}$ we have

$$-ig\,\bar{u}(\lambda_1', p_1)\gamma^\mu u(\lambda_1, p_1)\tau^a_{ij} = -2igp_1^\mu\delta_{\lambda_1'\lambda_1}\tau^a_{ij}.$$

This is called the **eikonal approximation** and it is valid whenever the gauge particle exchanged is 'soft' (i.e. all its components are small compared with the momentum of the emitting quark).

Remarkably, the eikonal approximation works not only for spin-$\frac{1}{2}$ quarks but for particles with any spin. If we had a scalar particle instead of a quark in Fig. 3.2 the upper vertex would be

$$-ig(2p_1 - q)^\mu\tau^a_{ij},$$

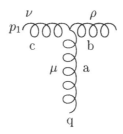

Fig. 3.3. Soft gluon emitted from a hard gluon.

which we approximate by $-2igp_1^\mu \tau_{ij}^a$. More importantly *it may be a gluon itself*, in which case the triple gluon vertex (see Fig. 3.3) is

$$ig\left[g^{\nu\rho}(2p_1 - q)^\mu + g^{\rho\mu}(2q - p_1)^\nu - g^{\mu\nu}(q + p_1)^\rho\right]T_{bc}^a$$

($T_{bc}^a = -if_{abc}$, where the f_{abc} are the structure constants of the gauge group, which we shall leave as $SU(N)$ so that the colour factor can be easily identified). Now neglecting q^μ and noting further that the incoming and outgoing gluons are on shell and therefore *transverse* (so that we may drop terms proportional to p_1^ν and $(p_1 - q)^\rho$) we once again end up with

$$2igp_1^\mu g^{\nu\rho}T_{bc}^a.$$

Thus, at lowest order, the amplitude for quark–quark scattering due to octet exchange is given by

$$\mathcal{A}_0^{(8)} = g^2 2p_1^\mu \frac{g_{\mu\nu}}{q^2} 2p_2^\nu \delta_{\lambda_1' \lambda_1} \delta_{\lambda_2' \lambda_2} G_0^{(8)} = 8\pi\alpha_s \frac{s}{t} \delta_{\lambda_1' \lambda_1} \delta_{\lambda_2' \lambda_2} G_0^{(8)},$$

(3.2)

where $\alpha_s = g^2/4\pi$ and $G_0^{(8)}$ is the colour factor for colour octet exchange, $\tau_{ij}^a \tau_{kl}^a$, which we shall subsequently write as $\tau^a \otimes \tau^a$.

We find it convenient to work in Feynman gauge although the amplitude is gauge invariant (the reader who tries to check this should remember that it will only work up to corrections of order t/s, since we have assumed that we may use the eikonal approximation).

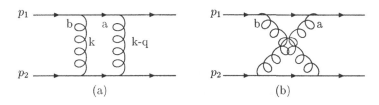

Fig. 3.4. Box and crossed box graphs.

3.2 Order α_s corrections

As explained in the preceding chapter, in leading $\ln s$ approximation we do *not* get contributions from one-loop graphs which contain corrections to propagators or to vertices, but only from the 'box' and 'crossed box' diagrams shown in Fig. 3.4 in which the loop integral depends on the centre-of-mass energy $\sqrt{s}$.[†]

Once again the contribution from Fig. 3.4(b) can be obtained from the contribution to Fig. 3.4(a) by crossing. However, in this case we not only have to interchange s and u (which introduces a minus sign since, in the Regge limit, $u \approx -s$) but also take into account a different colour factor. The colour factor for Fig. 3.4(a) is given by

$$G_a = (\tau^a \tau^b) \otimes (\tau^a \tau^b),$$

whereas the colour factor from Fig. 3.4(b) is

$$G_b = (\tau^a \tau^b) \otimes (\tau^b \tau^a).$$

Because crossing introduces a minus sign the total colour factor for octet exchange at the one-loop level is the difference between these two, i.e.

$$
\begin{aligned}
G_a - G_b &= (\tau^a \tau^b) \otimes [\tau^a, \tau^b] \\
&= i \frac{f_{abc}}{2} [\tau^a, \tau^b] \otimes \tau^c \\
&= \frac{i f_{abc} i f_{abd}}{2} \tau^d \otimes \tau^c = -\frac{N}{2} G_0^{(8)},
\end{aligned}
\tag{3.3}
$$

[†] This argument only works in a covariant gauge. In Coulomb or axial gauge in which an external vector is introduced, it is possible that vertex or self-energy corrections on upper (momentum p_1) lines can give rise to terms proportional to s through scalar products with the external vector which can have a component proportional to p_2. We confine ourselves to covariant gauges in this book.

where we have used the relation $f_{abc}f_{abd} = N\delta_{cd}$ for $SU(N)$, and $G_0^{(8)}$ is the colour factor for the tree diagram.

As in the preceding chapter, only Fig. 3.4(a) has an imaginary part so the imaginary part of the octet exchange amplitude at the one-loop level can be obtained using the Cutkosky rules and the tree level amplitude calculated in the preceding section (Eq.(3.2)), i.e.

$$\Im m A_{3.4a}^{(8)} = \frac{64\pi^2 \alpha_s^2}{2} \int d(P.S.^2) \left(\frac{s}{k^2}\right) \left(\frac{s}{(k-q)^2}\right) \delta_{\lambda_1' \lambda_1} \delta_{\lambda_2' \lambda_2} G_a.$$

In terms of the Sudakov variables $\rho, \lambda, \mathbf{k}$ of the momentum k^μ the two-body phase-space integration element may be written

$$d(P.S.^2) = \frac{s}{8\pi^2} d\rho \, d\lambda \, d^2\mathbf{k} \, \delta(-\lambda s - \mathbf{k}^2) \delta(\rho s - \mathbf{k}^2), \qquad (3.4)$$

where we have already made use of the inequalities $\rho, |\lambda| \ll 1$. In this approximation for which $-\rho \lambda s \ll \mathbf{k}^2$ we have

$$k^2 \approx -\mathbf{k}^2$$

and similarly

$$(k-q)^2 \approx -(\mathbf{k}-\mathbf{q})^2, \qquad (t = -\mathbf{q}^2),$$

hence

$$\Im m A_{3.4a}^{(8)} = 8\pi\alpha_s \frac{s}{t} \delta_{\lambda_1' \lambda_1} \delta_{\lambda_2' \lambda_2} G_a \frac{\alpha_s}{2\pi} \int d^2\mathbf{k} \frac{-\mathbf{q}^2}{\mathbf{k}^2(\mathbf{k}-\mathbf{q})^2}. \qquad (3.5)$$

Using $\ln(-s) = \ln s - i\pi$, this means that the real part is given by

$$\Re e A_{3.4a}^{(8)} = -8\pi\alpha_s \frac{s}{t} \delta_{\lambda_1' \lambda_1} \delta_{\lambda_2' \lambda_2} \ln(s/\mathbf{k}^2) G_a \frac{\alpha_s}{2\pi^2} \int d^2\mathbf{k} \frac{-\mathbf{q}^2}{\mathbf{k}^2(\mathbf{k}-\mathbf{q})^2}. \qquad (3.6)$$

Similarly the amplitude from Fig. 3.4(b) is

$$\Re e A_{3.4b}^{(8)} = -8\pi\alpha_s \frac{u}{t} \delta_{\lambda_1' \lambda_1} \delta_{\lambda_2' \lambda_2} \ln(-u/\mathbf{k}^2) G_b \frac{\alpha_s}{2\pi^2} \int d^2\mathbf{k} \frac{-\mathbf{q}^2}{\mathbf{k}^2(\mathbf{k}-\mathbf{q})^2}. \qquad (3.7)$$

(note that it is only the *sum* of these two which is actually octet exchange). Using $u \approx -s$ when $|t| \ll s$ and Eq.(3.3) we find that the complete one-loop amplitude in leading $\ln s$ approximation is given by

$$A_1^{(8)} = A_0^{(8)} \epsilon_G(t) \ln(s/\mathbf{k}^2), \qquad (3.8)$$

where

$$\epsilon_G(t) = \frac{N\alpha_s}{4\pi^2} \int d^2\mathbf{k} \frac{t}{\mathbf{k}^2(\mathbf{k}-\mathbf{q})^2} \quad (t = -\mathbf{q}^2). \tag{3.9}$$

The reader should notice that the integral on the right hand side of Eq.(3.9) is infra-red divergent. In the original work by Fadin, Kuraev & Lipatov (1976, 1977) and by Cheng & Lo (1976), great care was taken to regularize this divergence by breaking the gauge group spontaneously and including contributions from graphs in which there are Higgs bosons. For our purposes such rigour is not necessary. The infra-red divergence arises because the external quarks are on mass-shell. In the 'real world' this is not the case: the quarks are bound inside hadrons and off shell typically by an amount of the order of their average transverse momenta. Such an off-shellness provides a cut-off for the infra-red divergent integrals. Furthermore, it will turn out that the integral equation for the perturbative Pomeron is free from infra-red divergences. Therefore it is sufficient for us to leave ϵ_G in the form of Eq.(3.9), and it is to be understood that the infra-red divergence is to be regularized in some convenient way, introducing a scale which is expected to be of order Λ_{QCD}.

3.3 Order α_s^2 corrections

The two-loop corrections were performed independently by Tyburski (1976), Frankfurt & Sherman (1976) and by Lipatov (1976). We follow Lipatov's calculation closely.

As explained in the preceding chapter we do not get any contributions proportional to $\alpha_s^2 \ln^2 s$ from graphs which consist of vertex or self-energy insertions on the one-loop graphs considered in the last section. In order to obtain the imaginary part of the contribution in this order (in the leading $\ln s$ approximation) we need to consider the amplitude for a quark with momentum p_1 and a quark with momentum p_2 to scatter into a quark with momentum $p_1 - k_1$, a quark with momentum $p_2 + k_2$ and a gluon with momentum $k_1 - k_2$. Using Sudakov variables to parametrize the momenta k_1 and k_2:

$$k_i^\mu = \rho_i p_1^\mu + \lambda_i p_2^\mu + k_{i\perp}^\mu \quad (i = 1, 2),$$

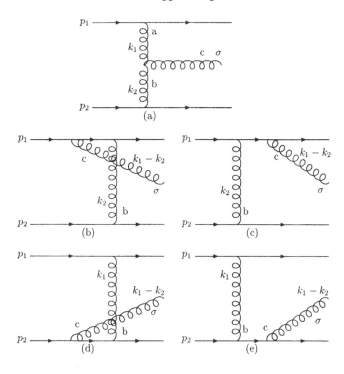

Fig. 3.5. Diagrams for the process qq → qq+g.

the leading logarithm contribution again comes from the region

$$1 \gg \rho_1 \gg \rho_2$$
$$1 \gg |\lambda_2| \gg |\lambda_1|$$

and the on-shell condition for the outgoing gluon becomes (in this approximation)

$$\rho_1\lambda_2 s = -(\mathbf{k_1} - \mathbf{k_2})^2,$$

so that $k_1^2 \approx k_{1\perp}^2 = -\mathbf{k_1^2}$ and $k_2^2 \approx k_{2\perp}^2 = -\mathbf{k_2^2}$. Once again the transverse momenta are both of the same magnitude ($\mathbf{k_1^2}, \mathbf{k_2^2}$ are both of order $\mathbf{k}^2$). The graphs for this process are shown in Fig. 3.5. We need these amplitudes in order to compute the 25 (two-loop) diagrams using the s-channel cutting rules.

The contribution from Fig. 3.5(a) (in Feynman gauge) in the

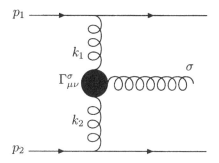

p_1

k_1

$\Gamma_{\mu\nu}^\sigma$ σ

k_2

p_2

Fig. 3.6. The effective non-local vertex.

relevant kinematic regime is

$$-ig^3 2s \left[\rho_1 p_1^\sigma + \lambda_2 p_2^\sigma - (k_1 + k_2)_\perp^\sigma \right] \frac{\delta_{\lambda_1' \lambda_1} \delta_{\lambda_2' \lambda_2}}{\mathbf{k_1^2 k_2^2}} f_{abc} \tau^a \otimes \tau^b.$$

The contributions from Fig. 3.5(b) and (c) in this regime are

$$-g^3 2s \frac{1}{\mathbf{k_2^2}} 2p_1^\sigma \left[\frac{\tau^b \tau^c}{(p_1 - k_1 + k_2)^2} + \frac{\tau^c \tau^b}{(p_1 - k_2)^2} \right] \otimes \tau^b \delta_{\lambda_1' \lambda_1} \delta_{\lambda_2' \lambda_2}.$$

Now $(p_1 - k_1 + k_2)^2 \approx s\lambda_2$ ($s \gg \mathbf{k}^2$) and similarly $(p_1 - k_2)^2 \approx -s\lambda_2$, so this contribution becomes

$$-g^3 2s \frac{2p_1^\sigma}{\mathbf{k_2^2} \lambda_2 s} [\tau^b, \tau^c] \otimes \tau^b \delta_{\lambda_1' \lambda_1} \delta_{\lambda_2' \lambda_2}$$

$$= -ig^3 2s \frac{2p_1^\sigma}{\mathbf{k_2^2} \lambda_2 s} f_{abc} \tau^a \otimes \tau^b \delta_{\lambda_1' \lambda_1} \delta_{\lambda_2' \lambda_2}.$$

Similarly the contributions from Fig. 3.5(d) and (e) are given by

$$-ig^3 2s \frac{2p_2^\sigma}{\mathbf{k_1^2} \rho_1 s} f_{abc} \tau^a \otimes \tau^b \delta_{\lambda_1' \lambda_1} \delta_{\lambda_2' \lambda_2}.$$

Although the contributions from Fig. 3.5(b) and (c) do not contain the denominator $\mathbf{k_1^2}$ and likewise the contributions from Fig. 3.5(d) and (e) do not contain the denominator $\mathbf{k_2^2}$, it is convenient to write all these contributions as though they all contained both of these denominators (multiplying by $\mathbf{k_1^2}$ or $\mathbf{k_2^2}$ where necessary) so that they may combined into an effective (left half of a) ladder, shown in Fig. 3.6.

The complete amplitude is

$$\mathcal{A}_{2\to 3}^{(8)\sigma} = -\frac{2ig^3 2 p_1^\mu p_2^\nu}{\mathbf{k}_1^2 \mathbf{k}_2^2} \delta_{\lambda_1' \lambda_1} \delta_{\lambda_2' \lambda_2} f_{abc} \tau^a \otimes \tau^b \Gamma_{\mu\nu}^\sigma(k_1, k_2), \quad (3.10)$$

where $\Gamma_{\mu\nu}^\sigma(k_1, k_2)$ is an effective (non-local) vertex given by

$$\Gamma_{\mu\nu}^\sigma(k_1, k_2) = \frac{2 p_{2\mu} p_{1\nu}}{s} \left[\left(\rho_1 + \frac{2\mathbf{k}_1^2}{\lambda_2 s} \right) p_1^\sigma \; + \; \left(\lambda_2 + \frac{2\mathbf{k}_2^2}{\rho_1 s} \right) p_2^\sigma \right.$$

$$\left. - (k_1 + k_2)_\perp^\sigma \right]. \qquad (3.11)$$

This vertex is said to be 'non-local' since it encodes the denominators of the propagators of Fig. 3.5(b–e). The dark blob in Fig. 3.6 represents the effective vertex.

We have been working in the Feynman gauge. Nevertheless the effective vertex $\Gamma_{\mu\nu}^\sigma(k_1, k_2)$ is gauge invariant. It can easily be shown to obey the Ward identity[†]

$$(k_1 - k_2)_\sigma \Gamma_{\mu\nu}^\sigma(k_1, k_2) = 0.$$

Individual graphs in Fig. 3.5 are gauge dependent, but the sum is gauge invariant.

It is fun to notice (and will be useful later when we consider higher order graphs) that we can exploit the gauge invariance in such a way that only the genuine ladder-type graph (Fig. 3.5(a)) contributes in leading logarithm order. If we remove the lower quark line from the graphs in Fig. 3.5 and write the amplitude as $\mathcal{M}_\tau^\sigma(k_1, k_2)$ (see Fig. 3.7), then since all but the bottom gluon are on mass-shell we have the Ward identity

$$k_2^\tau \mathcal{M}_\tau^\sigma(k_1, k_2) = 0. \qquad (3.12)$$

Now since the component of momenta proportional to p_2^τ in $\mathcal{M}_\tau^\sigma(k_1, k_2)$ is small we can neglect it and rewrite Eq.(3.12) as

$$\lambda_2 p_2^\tau \mathcal{M}_\tau^\sigma(k_1, k_2) + k_{2\perp}^\tau \mathcal{M}_\tau^\sigma(k_1, k_2) = 0.$$

In the eikonal approximation we have (reinstating the lower quark line) for the contributions from Fig. 3.5(a),(b) and (c)

$$\mathcal{A}_{abc}^{(8)\sigma} = 2 p_2^\tau \mathcal{M}_\tau^\sigma(k_1, k_2)$$

[†] Actually the Ward identity is only exact when the vertical gluon lines are on mass-shell. In fact these lines are off-shell by $\mathbf{k}_1^2$ and $\mathbf{k}_2^2$. However, since these (squared) transverse momenta are small compared with $\lambda_2 s$ or $\rho_1 s$ the identity is obeyed at the order to which we are working.

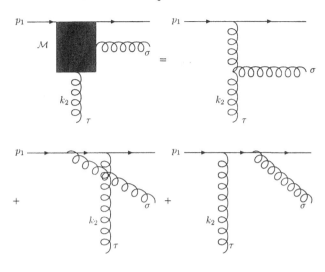

Fig. 3.7

(where we have dropped the colour factor and the coupling constant). This may be rewritten as

$$\mathcal{A}_{abc}^{(8)} = \frac{-2k_{2\perp}^\tau}{\lambda_2} \mathcal{M}_\tau^\sigma(k_1, k_2).\qquad(3.13)$$

Since, in the eikonal approximation, $\mathcal{M}_{\tau\sigma}$ has *no transverse components* from Fig. 3.5(b) and (c), it follows that Fig. 3.5(a) dominates.

We can of course play the same game by removing the upper quark line and write the corresponding Green function as $\mathcal{N}_\tau^\sigma(k_1, k_2)$. The amplitude for the graphs of Fig. 3.5(a),(d) and (e) can now be written

$$\mathcal{A}_{ade}^{(8)\sigma} = \frac{-2k_{1\perp}^\tau}{\rho_1} \mathcal{N}_\tau^\sigma(k_1, k_2)\qquad(3.14)$$

and once again it is only Fig. 3.5(a) that contributes at the leading logarithm level.

Now if we replace the eikonal insertion p_2^ν on the lower line by $-k_{2\perp}^\nu/\lambda_2$ and replace the eikonal insertion p_1^μ on the upper line by $-k_{1\perp}^\mu/\rho_1$ and consider only the dominant diagram, Fig. 3.5(a), we arrive at an alternative expression for $\mathcal{A}_{2\to3}^{(8)\sigma}$, i.e.

$$\mathcal{A}_{2\to3}^{(8)\sigma} = -\frac{4ig^3}{\mathbf{k}_1^2\mathbf{k}_2^2}\frac{k_{1\perp}^\mu k_{2\perp}^\nu}{\rho_1\lambda_2}\delta_{\lambda_1'\lambda_1}\delta_{\lambda_2'\lambda_2}f_{abc}\tau^a\otimes\tau^b$$
$$\left[-g_{\mu\nu}(k_1+k_2)^\sigma + g_\nu^\sigma(2k_2-k_1)_\mu + g_\mu^\sigma(2k_1-k_2)_\nu\right],\ (3.15)$$

where we have just used the ordinary triple gluon vertex. Writing k_1^μ and k_2^μ in terms of their Sudakov variables and making use of the inequalities $\rho_2 \ll \rho_1$ and $|\lambda_1| \ll |\lambda_2|$, this may be written as

$$\mathcal{A}_{2\to3}^{(8)\sigma} = \frac{2ig^3}{\mathbf{k}_1^2\mathbf{k}_2^2}\frac{1}{\rho_1\lambda_2}\delta_{\lambda_1'\lambda_1}\delta_{\lambda_2'\lambda_2}f_{abc}\tau^a\otimes\tau^b$$
$$\left\{\left[(\mathbf{k_1}-\mathbf{k_2})^2 - 2\mathbf{k}_1^2\right]\rho_1p_1^\sigma + \left[(\mathbf{k_1}-\mathbf{k_2})^2 - 2\mathbf{k}_2^2\right]\lambda_2p_2^\sigma\right.$$
$$-(\mathbf{k_1}-\mathbf{k_2})^2(k_1+k_2)_\perp^\sigma$$
$$\left.+(\mathbf{k}_1^2-\mathbf{k}_2^2)((\rho_1-\rho_2)p_1^\sigma + (\lambda_1-\lambda_2)p_2^\sigma + (k_1-k_2)_\perp^\sigma)\right\}.\ (3.16)$$

At first sight it does not appear that this works (i.e. we do not appear to be consistent with Eq.(3.11)). However, we note that the terms in the last line of Eq.(3.16) are proportional to $(k_1-k_2)^\sigma$. Since the outgoing gluon is on mass-shell it is transverse, and so terms proportional to $(k_1-k_2)^\sigma$ vanish when contracted with its polarization vector. These terms may therefore be dropped. Finally, using the on-shell condition for the outgoing gluon $(\mathbf{k_1}-\mathbf{k_2})^2 = -\rho_1\lambda_2 s$, we recover precisely Eq.(3.10).[†]

Returning now to the imaginary part of the octet exchange amplitude to order α_s^3, this is given by

$$\Im\mathrm{m}\mathcal{A}_2^{(8)} = \frac{-g_{\sigma\tau}}{2}\int d(P.S.^3)\mathcal{A}_{2\to3}^{(8)\sigma}(k_1,k_2)\mathcal{A}_{2\to3}^{\dagger(8)\tau}(k_1-q,k_2-q)$$
$$+ \text{ extra piece,}\ (3.17)$$

where the prefactor $-g_{\sigma\tau}$ arises from the sum over polarizations of the intermediate gluon and the 'extra piece' will be explained later in this section. We can take the components of q^μ to be transverse (more precisely the longitudinal components are negligible compared with $\rho_1\sqrt{s}$, $\lambda_2\sqrt{s}$).

[†] This is not a gauge choice (we are still working in Feynman gauge), but it is a trick which exploits the gauge invariance to reduce the effective ladder (Fig. 3.6) to the genuine ladder graph Fig. 3.5(a). It will be very useful in the next section.

We deal first with the colour factor which is

$$-f_{abc}f_{dec}(\tau^a\tau^d) \otimes (\tau^b\tau^e).$$

Anticipating that we shall be adding a contribution from the u-channel which will be equal and opposite to the s-channel contribution, but with τ^b and τ^e interchanged, we antisymmetrize in τ^b and τ^e. In other words we are 'sharing' the octet colour factor between the s-channel and u-channel contributions. We thus obtain

$$-\frac{1}{2}(f_{abc}f_{dec} - f_{aec}f_{cdb})(\tau^a\tau^d) \otimes (\tau^b\tau^e).$$

Making use of the Jacobi identity

$$f_{abc}f_{dec} + f_{aec}f_{bdc} + f_{adc}f_{ebc} = 0, \tag{3.18}$$

this becomes

$$-\frac{1}{2}f_{adc}f_{cbe}(\tau^a\tau^d) \otimes (\tau^b\tau^e).$$

The structure constants are antisymmetric in a, d and b, e, so we may replace the products of the colour matrices by commutators and obtain

$$\frac{1}{8}f_{adc}f_{adf}f_{cbe}f_{gbe}\tau^f \otimes \tau^g = \frac{N^2}{8}\tau^a \otimes \tau^a. \tag{3.19}$$

The phase-space integrand can now be written:

$$-\frac{1}{2}\mathcal{A}_{2\rightarrow3}^{(8)\sigma}(k_1, k_2)\mathcal{A}_{2\rightarrow3\sigma}^{\dagger(8)}(k_1 - q, k_2 - q)$$

$$= -\frac{g^6N^2}{16}\tau^a \otimes \tau^a\delta_{\lambda_1'\lambda_1}\delta_{\lambda_2'\lambda_2}\frac{16p_1^\mu p_2^\nu p_1^{\mu'} p_2^{\nu'}}{k_1^2k_2^2(k_1 - q)^2(k_2 - q)^2}$$

$$\times g_{\sigma\tau}\Gamma_{\mu\nu}^\sigma(k_1, k_2)\Gamma_{\mu'\nu'}^\tau(-(k_1 - q), -(k_2 - q)) \tag{3.20}$$

(recall that Hermitian conjugation requires the reversal of the direction of momentum in the right hand effective vertex). After a little algebra the right hand side of Eq.(3.20) becomes

$$-g^4\frac{N^2s}{4}\mathcal{A}_0^{(8)}\mathbf{q}^2\left[\frac{\mathbf{q}^2}{\mathbf{k_1^2 k_2^2(k_1 - q)^2(k_2 - q)^2}}\right.$$

$$\left.-\frac{1}{\mathbf{k_1^2(k_1 - k_2)^2(k_2 - q)^2}} - \frac{1}{\mathbf{k_2^2(k_1 - q)^2(k_1 - k_2)^2}}\right] \tag{3.21}$$

(the factor $(\mathbf{k_1} - \mathbf{k_2})^2$ in the denominators of the last two terms comes from replacing $p_1\lambda_2s$ by $-(\mathbf{k_1} - \mathbf{k_2})^2$). The three-body phase-space integral is

$$d(P.S.^3) = \frac{1}{(2\pi)^5} \left(\frac{s}{2}\right)^2 d\rho_1\, d\rho_2\, d\lambda_1\, d\lambda_2\, d^2\mathbf{k_1}\, d^2\mathbf{k_2}$$

$$\times\; \delta(-\lambda_1 s - \mathbf{k_1^2})\, \delta(\rho_2 s - \mathbf{k_2^2})\, \delta(-\rho_1\lambda_2 s - (\mathbf{k_1} - \mathbf{k_2})^2)$$

and, after performing the integration over $\rho_2, \lambda_1, \lambda_2$ (absorbing the delta functions), we obtain

$$\Im m \mathcal{A}_2^{(8)} = -\frac{N^2\alpha_s^2}{32\pi^3}\mathcal{A}_0^{(8)}\mathbf{q}^2 \int_{\mathbf{k}^2/s}^1 \frac{d\rho_1}{\rho_1} d^2\mathbf{k_1} d^2\mathbf{k_2}$$

$$\times \left[\frac{\mathbf{q}^2}{\mathbf{k_1^2}\mathbf{k_2^2}(\mathbf{k_1} - \mathbf{q})^2(\mathbf{k_2} - \mathbf{q})^2} - \frac{1}{\mathbf{k_1^2}(\mathbf{k_1} - \mathbf{k_2})^2(\mathbf{k_2} - \mathbf{q})^2} \right.$$

$$\left. - \frac{1}{\mathbf{k_2^2}(\mathbf{k_1} - \mathbf{q})^2(\mathbf{k_1} - \mathbf{k_2})^2} \right] + \text{ extra piece.} \qquad (3.22)$$

Some important cancellations have taken place to obtain the above expression. For example the terms in the product of the two effective vertices which give $\mathbf{k_1^2}, \mathbf{k_2^2}, (\mathbf{k_1} - \mathbf{q})^2$ or $(\mathbf{k_2} - \mathbf{q})^2$ in the numerator have cancelled. Had this not happened there would be integrals over the transverse momenta of the form

$$\int \frac{d^2\mathbf{k_1} d^2\mathbf{k_2}}{\mathbf{k_1^2}(\mathbf{k_1} - \mathbf{q})^2\mathbf{k_2^2}}, \qquad (3.23)$$

which is ultra-violet divergent. Of course the upper limit of the transverse momentum integrals is of order $\sqrt{s}$, so such integrals would not really diverge but would introduce a further factor of $\ln s$ (as well as the one we obtain from the integration over ρ_1). This would give an imaginary part proportional to $\ln^2 s$ and a real part proportional to $\ln^3 s$. Calculation of individual diagrams contributing to the order α_s^n correction to the tree amplitude do indeed contain terms proportional to $\alpha_s^n(\ln s)^{2n-1}$ but they cancel between graphs. In the case of QED this cancellation has been verified by explicit calculation up to four loops by McCoy & Wu (1976a–f).

The first term of Eq.(3.22) is encouraging since the integration over the transverse momenta factorizes and together with the logarithm from the integration over ρ_1 we obtain

$$-\frac{1}{2}\pi\epsilon_G^2(t)\ln\left(s/\mathbf{k}^2\right)\mathcal{A}_0^{(8)},$$

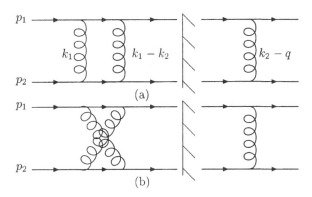

Fig. 3.8. Three gluon exchange graphs.

but the other two terms are not so nice. However, we have forgotten a contribution ('extra piece') coming from the diagrams shown in Fig. 3.8, which also contribute in leading ln s. Note that in these graphs the cut only goes through the quark lines. The contribution which arises when the cut also goes through the middle gluon line of Fig. 3.8(a) has been accounted for already in the interference between Fig. 3.5(c) and (d). There are two relevant contributions – one where there is one gluon exchanged on the right of the cut (shown in Fig. 3.8) and the other where there is one gluon exchanged on the left of the cut. Each of these gives a contribution to the imaginary part of $\mathcal{A}_2^{(8)}$ of

$$-\frac{4g^4}{2}\frac{N}{4}\int d(P.S.^2)\frac{s}{k_2^2}\epsilon_G(k_2^2)\ln(s/k^2)\frac{s}{(k_2-q)^2}\delta_{\lambda_1'\lambda_1}\delta_{\lambda_2'\lambda_2}$$

where we have made use of the result Eq.(3.8) for the amplitude on the left of the cut.

The colour factor, $N/4$, is obtained in the same way as in the preceding section (projecting the colour octet exchange part). Now, from Eq.(3.9)

$$\epsilon_G(k_2^2) = -\frac{N\alpha_s}{4\pi^2}\int d^2k_1\frac{k_2^2}{(k_2-k_1)^2k_1^2},$$

and integrating over λ_1, ρ_1 using the two-body phase-space expression Eq.(3.4) we obtain a contribution of

$$-\frac{N^2\alpha_s^2}{32\pi^3}\mathcal{A}_0^{(8)}\mathbf{q}^2\int\frac{d^2\mathbf{k}_1 d^2\mathbf{k}_2}{\mathbf{k}_1^2(\mathbf{k}_1-\mathbf{k}_2)^2(\mathbf{k}_2-\mathbf{q})^2}\ln\left(s/\mathbf{k}^2\right). \qquad (3.24)$$

Together with the contribution from the graphs with one gluon to the left of the cut this exactly cancels the 'unwanted' parts of Eq.(3.22) and we are left with an imaginary part:

$$\Im m\mathcal{A}_2^{(8)}=-\frac{1}{2}\epsilon_G^2(t)\pi\ln\left(s/\mathbf{k}^2\right)\mathcal{A}_0^{(8)}. \qquad (3.25)$$

The corresponding real part is

$$\Re e\mathcal{A}_2^{(8)}=\frac{1}{4}\epsilon_G^2(t)\ln^2(s/\mathbf{k}^2)\mathcal{A}_0^{(8)}. \qquad (3.26)$$

We obtain a similar contribution from the crossed diagrams with s replaced by u (and a further sign from the colour factor). Thus up to order α_s^2 we have a colour octet amplitude given in leading $\ln s$ approximation by

$$\mathcal{A}_0^{(8)}\left(1+\epsilon_G(t)\ln\left(s/\mathbf{k}^2\right)+\frac{1}{2}\epsilon_G^2(t)\ln^2(s/\mathbf{k}^2)+\cdots\right). \qquad (3.27)$$

It is tempting to speculate that these are the first three terms in the expansion of $\mathcal{A}_0^{(8)}s^{\epsilon_G(t)}$. Cheng & Lo (1976) showed that this trend continues up to three loops. In the following section we shall show that it continues to work to all orders of perturbation theory. It is worth emphasizing at this point that the remarkable cancellation between the 'extra piece' from graphs in which three gluons are exchanged between the quarks and the unwanted contribution from the graphs in which three lines are cut depends crucially on the colour factors working out just right. Whereas it works for colour octet exchange, it fails for other channels, particularly for the colour singlet exchange channel which we shall need in order to study the Pomeron.

3.4 The $2\to(n+2)$ amplitude at the tree level

It was explained earlier in this chapter that the eikonal approximation is independent of the spin of the high energy particle which emits the soft gluon. We may therefore replace the quark lines in Fig. 3.5 by gluons themselves. The eikonal approximation is still valid because of the strong ordering of the momenta. The effective vertex (Eq.(3.11)) is the vertex obtained by adding a gluon with

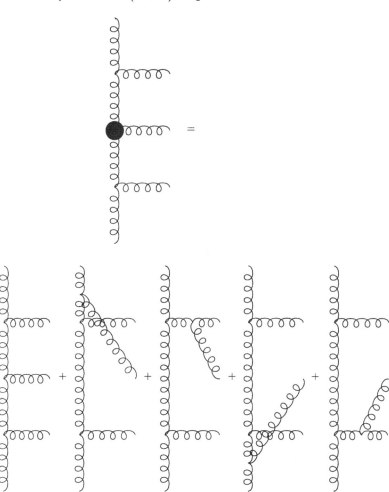

Fig. 3.9

momentum $(k_1 - k_2)^\mu$ to all the gluon lines in a gluon–gluon scattering amplitude with colour octet exchange, as shown in Fig. 3.9.

One might guess that adding more gluons generates more factors of the effective vertices (together with extra propagators for the vertical gluons), giving rise to (the left half of) an n-rung ladder with effective vertices, Γ, at each intersection, so that the amplitude for two quarks to scatter into two quarks and n gluons

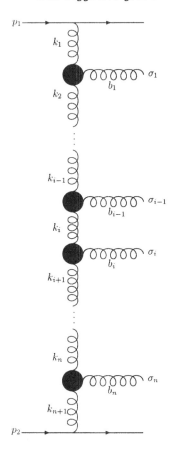

Fig. 3.10. Tree amplitude for two quarks to two quarks plus n gluons.

with octet colour exchange is shown in Fig. 3.10. It turns out that, in the kinematic regime that we are interested in, namely, where the ith emitted gluon has momentum $(k_i - k_{i+1})^\mu$ with Sudakov variables for k_i^μ and k_{i+1}^μ obeying the inequalities

$$1 \gg \rho_i \gg \rho_{i+1} \gg \mathbf{k}^2/s$$
$$1 \gg |\lambda_{i+1}| \gg |\lambda_i| \gg \mathbf{k}^2/s,$$

this guess is correct. Thus in this limit the amplitude for two quarks to scatter into two quarks and n gluons with colour octet

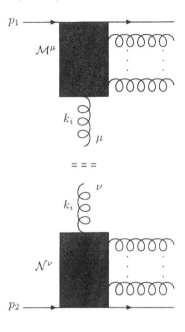

Fig. 3.11

exchange is given by

$$\mathcal{A}_{2\to(n+2)}^{(8)\sigma_1\cdots\sigma_n} = i2s\,(g)^{n+2}\,\delta_{\lambda_1'\lambda_1}\delta_{\lambda_2'\lambda_2}G_n^{(8)}$$

$$\times \frac{i}{\mathbf{k}_1^2}\prod_{i=1}^{n}\left\{\frac{2p_1^{\mu_i}p_2^{\nu_{i+1}}}{s}\Gamma_{\mu_i\nu_{i+1}}^{\sigma_i}(k_i,k_{i+1})\frac{i}{\mathbf{k}_{i+1}^2}\right\},(3.28)$$

where the colour factor $G_n^{(8)}$ (for gluons with colours b_1 to b_n) is

$$G_n^{(8)}(b_1,\cdots b_n) = \prod_{i=1}^{n} f_{a_i a_{i+1} b_i}\,\tau^{a_1}\otimes\tau^{a_{n+1}}.\qquad (3.29)$$

A rather elegant derivation of Eq.(3.28) is given by Gribov, Levin & Ryskin (1983). We reproduce their derivation here. The reader who is prepared to accept Eq.(3.28) on trust may skip to the next section.

Consider the amplitude for two quarks to scatter into two quarks plus n gluons. As described in the last section if we cut the ith vertical gluon, whose momentum is k_i, the amplitude separates into an upper part $\mathcal{M}_\mu(p_1, k_1, \cdots k_i)$ and a lower part

$\mathcal{N}_\nu(p_2, k_i, \cdots k_n)$ (see Fig. 3.11). Since all but the cut gluon line are on shell, these Green functions obey the Ward identities

$$k_i^\mu \mathcal{M}_\mu(p_1, k_1, \cdots k_i) = 0 \qquad (3.30)$$

$$k_i^\nu \mathcal{N}_\nu(p_2, k_i \cdots k_n) = 0. \qquad (3.31)$$

The largest momentum in the amplitude $\mathcal{M}_\mu$ is p_1 and so the largest part of $\mathcal{M}_\mu$ will be proportional to p_1^μ. Likewise, the largest part of $\mathcal{N}_\nu$ will be proportional to p_2^ν. Therefore in leading logarithm approximation we may rewrite Eqs.(3.30) and (3.31) as

$$k_{i\perp}^\mu \mathcal{M}_\mu(p_1, k_1, \cdots k_i) = -\lambda_i p_2^\mu \mathcal{M}_\mu(p_1, k_1, \cdots k_i)$$

$$k_{i\perp}^\nu \mathcal{N}_\nu(p_2, k_i, \cdots k_n) = -\rho_i p_1^\nu \mathcal{N}_\nu(p_2, k_i, \cdots k_n).$$

This means that we may replace the numerator of the cut gluon propagator by

$$\frac{2 k_{i\perp}^\mu k_{i\perp}^\nu}{\rho_i \lambda_i s}. \qquad (3.32)$$

We can cut any of the intermediate vertical gluon lines and perform the same manipulations. Therefore, we end up with an amplitude which can be obtained from (the left half of) a genuine uncrossed ladder in which the numerator of the vertical gluon lines is replaced by the expression (3.32). We associate a factor of $\sqrt{(2/s)} k_{i\perp}^\mu / \lambda_i$ with the vertex at the top of the ith vertical gluon and a factor of $\sqrt{(2/s)} k_{i\perp}^\nu / \rho_i$ with the vertex at the bottom of the ith vertical gluon. The amplitude thus becomes

$$\mathcal{A}_{2\to(n+2)}^{(8)\sigma_1\cdots\sigma_n} = 2isg^2 \frac{i}{\mathbf{k}_1^2} \delta_{\lambda_1'\lambda_1} \delta_{\lambda_2'\lambda_2} G_n^{(8)} \prod_{i=1}^n \frac{ig}{\mathbf{k}_{i+1}^2} \frac{2 k_{i\perp}^{\mu_i} k_{i+1\perp}^{\nu_i}}{\lambda_{i+1}\rho_i s}$$

$$\times \left[g_{\mu_i\nu_i}(-k_i - k_{i+1})^{\sigma_i} + g_{\mu_i}^{\sigma_i}(2k_i - k_{i+1})_{\nu_i} \right.$$

$$\left. + g_{\nu_i}^{\sigma_i}(2k_{i+1} - k_i)_{\mu_i} \right]. \qquad (3.33)$$

We showed in the last section that

$$\frac{2 k_{i\perp}^{\mu_i} k_{i+1\perp}^{\nu_i}}{\lambda_{i+1}\rho_i s} \left[g_{\mu_i\nu_i}(-k_i - k_{i+1})^{\sigma_i} + g_{\mu_i}^{\sigma_i}(2k_i - k_{i+1})_{\nu_i} \right.$$

$$\left. + g_{\nu_i}^{\sigma_i}(2k_{i+1} - k_i)_{\mu_i} \right] = \frac{2 p_2^{\mu_i} p_1^{\nu_i}}{s} \Gamma_{\mu_i\nu_i}^{\sigma_i}(k_i, k_{i+1}), \qquad (3.34)$$

plus terms proportional to $(k_i - k_{i+1})^{\sigma_i}$, which vanish because the outgoing gluon is on shell and therefore transverse. The result, Eq.(3.28), then follows.

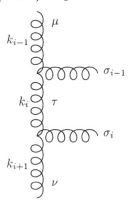

Fig. 3.12

We now need to show that, *using this gauge technique*, the diagrams which are not of the form of uncrossed ladders give contributions which are suppressed by at least one power of ρ_i/ρ_{i+1} and therefore will only contribute to sub-leading logarithm terms when the (phase-space) integrals over all ρ_i s are performed.

First of all let us look at the ith section of (the left half of) the uncrossed ladder (Fig. 3.12). The contribution from this section is proportional to the two effective vertices

$$\Gamma_{\mu\tau}^{\sigma_{i-1}}\Gamma_{\nu}^{\tau\sigma_i}.$$

The leading contribution proportional to $p_1^{\sigma_{i-1}}p_2^{\sigma_i}$ is

$$\sim \frac{2p_{2\mu}p_{1\nu}}{s}\rho_{i-1}\lambda_{i+1}p_1^{\sigma_{i-1}}p_2^{\sigma_i} \tag{3.35}$$

and the contribution proportional to $k_{i-1\perp}^{\sigma_{i-1}}k_{i\perp}^{\sigma_i}$ is

$$\sim \frac{2p_{2\mu}p_{1\nu}}{s}k_{i-1\perp}^{\sigma_{i-1}}k_{i+1\perp}^{\sigma_i}. \tag{3.36}$$

Since cross-rung graphs involve sections of the ladder where the momenta of incoming and outgoing gluons at the ith vertex are not simply k_i, k_{i+1} (see Fig. 3.13) we need to generalize the formula (3.34) for the case where the upper line entering the vertex has momentum k_i^μ and the lower line has momentum k_j^μ. This leads to

$$\frac{2k_{i\perp}^\mu k_{j\perp}^\nu}{\rho_i\lambda_j s}\left[-g_{\mu\nu}(k_i + k_j)^\sigma + g_\mu^\sigma(2k_i - k_j)_\nu + g_\nu^\sigma(2k_j - k_i)_\mu\right],$$

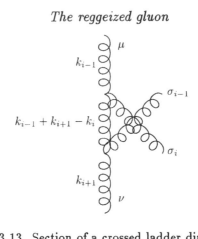

Fig. 3.13. Section of a crossed ladder diagram

which is

$$\sim \frac{2}{\rho_i \lambda_j s} \mathbf{k}^2 \left[\rho_i p_1^\sigma + \lambda_j p_2^\sigma + (k_i + k_j)_\perp^\sigma \right].$$

Since $\rho_{j-1} \lambda_j$ is of order $\mathbf{k}^2/s$ from the on-shell condition of the jth outgoing gluon, we have a contribution of order

$$2 \frac{\rho_{j-1} s}{\rho_i} \left[\rho_i p_1^\sigma + \lambda_j p_2^\sigma + (k_i + k_j)_\perp^\sigma \right].$$

Now imagine a section of a crossed-rung ladder (shown in Fig. 3.13) where the middle vertical line has momentum $(k_{i-1} + k_{i+1} - k_i)$, giving rise to a denominator from its propagator which is approximately equal to $\rho_{i-1} \lambda_{i+1} s$. The two vertices have a component proportional to $p_1^{\sigma_{i-1}} p_2^{\sigma_i}$ which is of order

$$\frac{\rho_i}{\rho_{i-1}} \frac{\rho_i}{\rho_{i-1}} \frac{2 p_2^\mu p_1^\nu}{s} \rho_{i-1} \lambda_{i+1} p_1^{\sigma_{i-1}} p_2^{\sigma_i}. \tag{3.37}$$

The factors of ρ_i in the numerator of Eq.(3.37) occur because $\lambda_j \approx \lambda_{i+1}$. This is true at both vertices because $|\lambda_{i+1}| \gg |\lambda_i|$ (or $|\lambda_{i-1}|$). Using the on-shell conditions we may therefore replace ρ_{j-1} by ρ_i to arrive at Eq.(3.37).

Since $\rho_i \ll \rho_{i-1}$, expression (3.37) is much smaller than the uncrossed ladder product of two effective vertices, expression (3.35). In addition to this suppression the denominator from the propagator of the intermediate line is much larger than $\mathbf{k}_i^2$, which is what we obtain from the section of the ladder shown in Fig. 3.12. Thus there is a double suppression of the crossed ladder diagram. If we cross more rungs we get an even greater suppression.

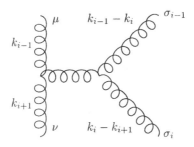

Fig. 3.14

Let us now consider a section of a graph in which two of the out-going gluons meet at a point. Such a section of a graph involving the triple gluon vertex is shown in Fig. 3.14. Again contracting the left hand triple gluon vertex with $k^{\mu}_{i-1\perp} k^{\nu}_{i+1\perp}$, we obtain a term proportional to $k^{\sigma_{i-1}}_{i-1\perp} k^{\sigma_i}_{i\perp}$ which is of order

$$\frac{2p_1^{\mu} p_2^{\nu}}{s} \mathbf{k}^2 \rho_{i-1} \lambda_{i+1} s k^{\sigma_{i-1}}_{i-1\perp} k^{\sigma_i}_{i\perp}$$

and again using the fact that $\rho_i \lambda_{i+1}$ is of order $\mathbf{k}^2/s$ this is of order

$$\frac{2p_1^{\mu} p_2^{\nu}}{s} \frac{\rho_i}{\rho_{i-1}} k^{\sigma_{i-1}}_{i-1\perp} k^{\sigma_i}_{i\perp},$$

which is suppressed relative to the equivalent term from the un-crossed ladder (expression (3.36)) by a factor of ρ_i/ρ_{i-1}. In addition to this the internal gluon propagator has a denominator which is again much larger than $\mathbf{k}^2$, so we get a double suppression.

From the four-point gluon vertex we get a section of a graph shown in Fig. 3.15. Once again the contribution from the vertex has a term proportional to $k^{\sigma_{i-1}}_{i-1\perp} k^{\sigma_i}_{i\perp}$ which is of order

$$\frac{1}{\rho_{i-1} \lambda_{i+1} s} k^{\sigma_{i-1}}_{i-1\perp} k^{\sigma_i}_{i\perp}$$

and we are missing a propagator factor of $\mathbf{k}_i^2$ present in the section of the graph shown in Fig. 3.12. Thus this graph also gives a contribution which is suppressed relative to the uncrossed ladder contribution by a factor of order ρ_i/ρ_{i-1}.

Comparison of other components of the tensor structure (e.g. terms proportional to $p_1^{\sigma_{i-1}} p_2^{\sigma_i}$) yield similar suppression factors.

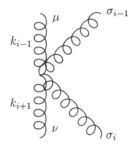

Fig. 3.15

This, then, completes the proof that the amplitude for two quarks to scatter into two quarks and n gluons via colour octet exchange is given by Eq.(3.28) in the kinematic region that leads to leading logarithms.

3.5 Absence of fermion loops

We have so far only considered outgoing gluons in addition to the two quarks present in the initial state. In principle we must also consider the production of extra fermion–antifermion pairs, since such amplitudes must be included in the dispersion relation for the imaginary part of the elastic scattering amplitude. However, these also turn out to be suppressed and do not contribute in leading logarithm approximation. The essential reason for this is that a fermion exchanged in the t-channel gives an s-dependence which is lower than that of an exchanged vector particle due to the fact that the fermion has spin $\frac{1}{2}$.

Looking at this in more detail, in Fig. 3.16(a) we display a section of a ladder in which two of the gluons are replaced by a fermion–antifermion pair. Once again we may use the gauge technique to replace the factor of p_1^μ from the upper gluon by $\sqrt{(2/s)}k_{i-1\perp}^\mu/\rho_{i-1}$ and the factor of p_2^ν from the lower gluon line by $\sqrt{(2/s)}k_{i+1\perp}^\nu/\lambda_{i+1}$. Having done this the contribution from the section shown in Fig. 3.16(a) contains terms proportional to

$$\frac{1}{\rho_{i-1}\lambda_{i+1}s}\bar{u}(k_{i-1}-k_i)\gamma\cdot k_{i-1\perp}\gamma\cdot k_i\gamma\cdot k_{i+1\perp}u(k_{i+1}-k_i).$$

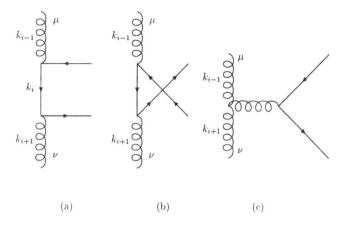

Fig. 3.16. Section of a ladder with a fermion loop.

This is of order

$$\frac{\mathbf{k}^2}{\rho_{i-1}\lambda_{i+1}s}\{k_i \cdot k_{i-1}, k_i \cdot k_{i+1}, k_i^2\}.$$

Now all the scalar products inside the braces are of order $\mathbf{k}^2$ ($\rho_{i-1}\lambda_i$ and $\rho_i\lambda_{i+1}$ are both of order $\mathbf{k}^2/s$), and the factor outside the braces is of order ρ_i/ρ_{i-1}. Thus we obtain a contribution which is suppressed by ρ_i/ρ_{i-1} compared with a typical term from the gluon ladder.

Examination of the graphs shown in Fig. 3.16(b) and (c) also give a similar suppression factor, although in these cases it is due to the presence of a hard fermion or gluon propagator.

Thus we see that it is sufficient to neglect fermion–antifermion pair production in the final state in order to obtain the imaginary part of the elastic amplitude to leading logarithm order.

3.6 Ladders within ladders

We now have an expression for the tree level amplitude for two quarks to scatter to two quarks and n gluons, which when multiplied by the conjugate amplitude and integrated over phase space contributes to the imaginary part of the 'reggeized gluon' amplitude. We now consider loop corrections. A strong hint on how to handle these is given by the fact that it was necessary to consider the graphs of Fig. 3.8 at the two-loop level in order to obtain a result that looks like the first three terms of the expansion of the required reggeized form. The subgraphs on the left of the cut in Fig. 3.8 may be viewed as the beginning of an expansion of a ladder itself.

The upshot of all this is that the imaginary part of the octet exchange amplitude in leading $\ln s$ is

a superposition of n-rung ladders with effective vertices at each rung, whose vertical lines are a superposition of n-rung ladders with effective vertices at each rung, whose vertical lines are a superposition of n-rung ladders with effective vertices at each rung, whose vertical lines are a superposition of n-rung ladders with effective vertices at each rung whose vertical lines are a superposition of n-rung ladders with effective vertices at each rung ...

(n runs from 0 to ∞). The effect of these ladders is to 'reggeize' the gluon, i.e. if we consider the ith section of the ladder (see Fig. 3.12) the square of the centre-of-mass energy coming into this section is

$$\hat{s}_i = (k_{i-1} - k_{i+1})^2 \approx -\rho_{i-1}\lambda_{i+1}s = \frac{\rho_{i-1}}{\rho_i}(\mathbf{k_i} - \mathbf{k_{i+1}})^2 \quad (3.38)$$

(where in the last step we have used the on-shell condition for the ith outgoing gluon).

The reggeization simply means that the propagator of the ith vertical gluon (in Feynman gauge) is replaced by

$$\tilde{D}_{\mu\nu}(\hat{s}_i, k_i^2) = \frac{ig_{\mu\nu}}{\mathbf{k_i^2}} \left(\frac{\hat{s}_i}{\mathbf{k}^2}\right)^{\epsilon_G(k_i^2)}. \quad (3.39)$$

Since all the transverse momenta are of the same order we may replace $(\mathbf{k_i} - \mathbf{k_{i+1}})^2$ in Eq.(3.38) by a typical squared transverse momentum, $\mathbf{k}^2$, and rewrite this as

$$\tilde{D}_{\mu\nu}(\hat{s}_i, k_i^2) = \frac{ig_{\mu\nu}}{\mathbf{k_i^2}} \left(\frac{\rho_{i-1}}{\rho_i}\right)^{\epsilon_G(k_i^2)}. \quad (3.40)$$

We shall establish the validity of this proposition by a 'bootstrap' method. Encouraged by the results of the first few orders in perturbation theory, we shall assume that Eq.(3.40) is true. This will enable us to establish an integral equation for the (Mellin transform of) the amplitude for colour octet exchange. The integral equation has a solution in which the Mellin transform has a pole at $\omega = \epsilon_G(t)$, (implying an $s^{\alpha_G(t)}$ behaviour) and this justifies the assumption of reggeization used to establish the integral equation in the first place. It demonstrates the self-consistency of the proposition and, together with the results of the first few orders in perturbation theory, provides an inductive derivation of reggeization valid to all orders in perturbation theory.

The horizontal gluon rungs are attached to the vertical lines via effective vertices $\Gamma^\sigma_{\mu\nu}(k_i, k_{i+1})$ and so the amplitude for two quarks to scatter into two quarks plus n gluons via colour octet exchange becomes

$$
\begin{aligned}
\mathcal{A}^{(8)\sigma_1\cdots\sigma_n}_{2\to(n+2)} &= i2sg^{n+2}\delta_{\lambda'_1\lambda_1}\delta_{\lambda'_2\lambda_2}G^{(8)}_n \frac{i}{\mathbf{k}_1^2}\left(\frac{1}{\rho_1}\right)^{\epsilon_G(k_1^2)} \\
&\times \prod_{i=1}^{n}\frac{2p_1^{\mu_i}p_2^{\nu_{i+1}}}{s}\Gamma^{\sigma_i}_{\mu_i\nu_{i+1}}(k_i, k_{i+1})\frac{i}{\mathbf{k}_{i+1}^2}\left(\frac{\rho_i}{\rho_{i+1}}\right)^{\epsilon_G(k_{i+1}^2)}.
\end{aligned}
\tag{3.41}
$$

In actual fact this is the **multi-Regge exchange** amplitude for the $2 \to 2+n$ amplitude via the exchange of $n+1$ reggeized particles with Regge trajectory $\alpha_G(k_i^2)$. This can be established using techniques of Regge theory, exploiting unitarity in all possible final state sub-channels. This long calculation was performed by Bartels (1975) and is outlined by Lipatov (1989) and we refer the reader to the literature for further details. We shall now proceed to demonstrate the self-consistency of the reggeization *ansatz*.

3.7 The integral equation

The imaginary part of the octet exchange amplitude is given by (see Fig. 3.17 in which a dash on the vertical gluon lines indicates that they are reggeized gluons)

$$
\begin{aligned}
\Im m\mathcal{A}^{(8)}(s,t) &= \frac{1}{2}\sum_{n=0}^{\infty}(-1)^n\int d(P.S.^{(n+2)})\left(\mathcal{A}^{(8)\sigma_1\cdots\sigma_n}_{2\to(n+2)}(k_1,\cdots k_n)\right. \\
&\times \left.\mathcal{A}^{(8)\dagger}_{2\to(n+2)\sigma_1\cdots\sigma_n}(k_1-q,\cdots k_n-q)\right),
\end{aligned}
\tag{3.42}
$$

and $d(P.S.^{(n+2)})$ is the $(n+2)$-body phase space given in Eq.(2.27) in the preceding chapter.

The colour factor is readily calculated using repetitions of the Jacobi identity (Eq.(3.18)) as was done to obtain the colour factor

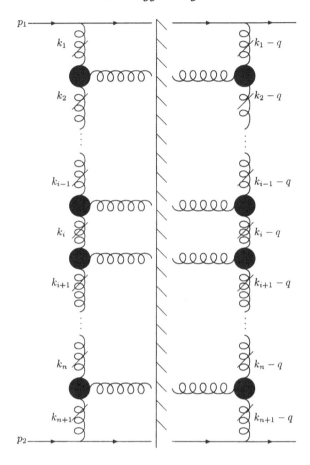

Fig. 3.17. n-rung ladder contribution to imaginary part of amplitude. The dashes on the vertical gluon lines indicate that they are reggeized gluons.

at the two-loop level (Eq.(3.19)). The result is

$$G_n^{(8)}(b_1, \cdots b_n) \times G_n^{(8)}(b_1, \cdots b_n) = \left(\frac{N}{2}\right)^n \frac{N}{4} \tau^a \otimes \tau^a.$$

Performing the contractions of the effective vertices we obtain

$$\Im m \mathcal{A}^{(8)}(s,t) = \sum_{n=0}^{\infty} \int d(P.S.^{(n+2)})$$

$$\times \frac{N}{4} \mathcal{A}_0^{(8)}(s,t)(-N)^n \frac{g^2 s q^2}{k_1^2 (k_1 - q)^2} \left(\frac{1}{\rho_1}\right)^{\epsilon_G(k_1^2) + \epsilon_G((k_1 - q)^2)}$$

$$\times \prod_{i=1}^{n} \left[\frac{g^2}{k_{i+1}^2 (k_{i+1} - q)^2} \left(q^2 - \frac{k_i^2 (k_{i+1} - q)^2 + (k_i - q)^2 k_{i+1}^2}{(k_i - k_{i+1})^2} \right) \right.$$

$$\left. \times \left(\frac{\rho_i}{\rho_{i+1}}\right)^{\epsilon_G(k_{i+1})^2 + \epsilon_G((k_{i+1} - q)^2)} \right] \tag{3.43}$$

(for $n = 0$ the product in Eq.(3.43) is replaced by 1). The reader can check that, apart from the reggeization factors, the $n = 0$ and $n = 1$ terms correspond to (the s-channel contributions to) $\Im m \mathcal{A}_1^{(8)}$ (Eq.(3.5)) and $\Im m \mathcal{A}_2^{(8)}$ (Eq.(3.22)) respectively.

We note that the integrations over the ρ_i are nested and the best way to unravel them is to take the Mellin transform and make use of the convolution formula, Eq.(A.2.7). To this end we define a quantity $\mathcal{F}^{(8)}(\omega, \mathbf{k}, \mathbf{q})$ by

$$\int \left(\frac{\Im m \mathcal{A}^{(8)}(s,t)}{\mathcal{A}_0^{(8)}(s,t)}\right) \left(\frac{s}{k^2}\right)^{-\omega - 1} d\left(\frac{s}{k^2}\right) = \int \frac{d^2 k}{k^2 (k - q)^2} \mathcal{F}^{(8)}(\omega, \mathbf{k}, \mathbf{q}). \tag{3.44}$$

The Mellin transform and integration over the ρ_i then leaves us with

$$\mathcal{F}^{(8)}(\omega, \mathbf{k}, \mathbf{q}) = \sum_{n=0}^{\infty} \frac{\pi}{2} \left(\frac{\alpha_s N}{4\pi^2}\right)^{n+1} (-1)^n$$

$$\times \frac{q^2}{(\omega - \epsilon_G(-k^2) - \epsilon_G(-(k-q)^2))} \int d^2 k_{n+1}$$

$$\times \prod_{i=1}^{n} \left[\int \frac{d^2 k_i}{k_i^2 (k_i - q)^2} \frac{1}{(\omega - \epsilon_G(-k_i^2) - \epsilon_G(-(k_i - q)^2))} \right.$$

$$\left. \times \left(q^2 - \frac{k_i^2 (k_{i+1} - q)^2 + k_{i+1}^2 (k_i - q)^2}{(k_i - k_{i+1})^2} \right) \right] \delta^2(k - k_{n+1}). \tag{3.45}$$

This sum of all ladders is most easily treated by obtaining an integral equation for $\mathcal{F}^{(8)}(\omega, \mathbf{k}, \mathbf{q})$. This integral equation, shown

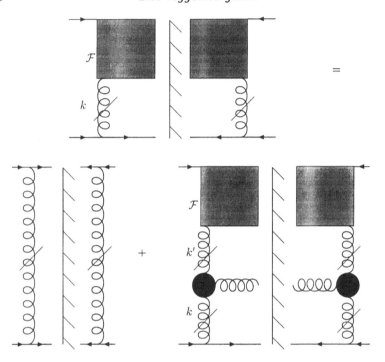

Fig. 3.18. Integral equation for imaginary part of the octet exchange amplitude.

diagrammatically in Fig. 3.18 (where again a dash on a gluon line indicates that it is a reggeized gluon), is

$$\mathcal{F}^{(8)}(\omega, \mathbf{k}, \mathbf{q}) = \frac{\pi}{2} \frac{\alpha_s N}{4\pi^2} \frac{\mathbf{q}^2}{(\omega - \epsilon_G(-\mathbf{k}^2) - \epsilon_G(-(\mathbf{k} - \mathbf{q})^2))}$$
$$- \frac{\alpha_s N}{4\pi^2} \int d^2\mathbf{k}' \frac{\mathcal{F}^{(8)}(\omega, \mathbf{k}', \mathbf{q})}{(\omega - \epsilon_G(-\mathbf{k}^2) - \epsilon_G(-(\mathbf{k} - \mathbf{q})^2))}$$
$$\times \frac{1}{\mathbf{k}'^2(\mathbf{k}' - \mathbf{q})^2} \left(\mathbf{q}^2 - \frac{\mathbf{k}^2(\mathbf{k}' - \mathbf{q})^2 + \mathbf{k}'^2(\mathbf{k} - \mathbf{q})^2}{(\mathbf{k} - \mathbf{k}')^2} \right). \quad (3.46)$$

The first term represents the exchange of two reggeized gluons with no rungs on the ladder. The second term represents the effect of adding a rung which couples with effective vertices to the vertical lines, which are themselves reggeized, and serves to build up the sum of all ladders (as discussed in Section 2.5).

This rather forbidding looking equation in actual fact has a rather simple solution in which $\mathcal{F}^{(8)}(\omega, \mathbf{k}, \mathbf{q})$ is independent of $\mathbf{k}$. To see this we multiply by $(\omega - \epsilon_G(-\mathbf{k}^2) - \epsilon_G(-(\mathbf{k} - \mathbf{q})^2))$ to obtain

$$(\omega - \epsilon_G(-\mathbf{k}^2) - \epsilon_G(-(\mathbf{k} - \mathbf{q})^2))\mathcal{F}^{(8)}(\omega, \mathbf{k}, \mathbf{q})$$

$$= \frac{\pi}{2} \frac{\alpha_s N \mathbf{q}^2}{4\pi^2} - \frac{\alpha_s N}{4\pi^2} \int d^2\mathbf{k}' \, \mathcal{F}^{(8)}(\omega, \mathbf{k}', \mathbf{q})$$

$$\times \left(\frac{\mathbf{q}^2}{\mathbf{k}'^2(\mathbf{k}' - \mathbf{q})^2} - \frac{\mathbf{k}^2}{\mathbf{k}'^2(\mathbf{k} - \mathbf{k}')^2} - \frac{(\mathbf{k} - \mathbf{q})^2}{(\mathbf{k}' - \mathbf{q})^2(\mathbf{k} - \mathbf{k}')^2} \right). \quad (3.47)$$

Now we note that

$$\epsilon_G(-\mathbf{k}^2) = -\frac{\alpha_s N}{4\pi^2} \int d^2\mathbf{k}' \frac{\mathbf{k}^2}{\mathbf{k}'^2(\mathbf{k} - \mathbf{k}')^2} \quad (3.48)$$

$$\epsilon_G(-(\mathbf{k} - \mathbf{q})^2) = -\frac{\alpha_s N}{4\pi^2} \int d^2\mathbf{k}' \frac{(\mathbf{k} - \mathbf{q})^2}{(\mathbf{k}' - \mathbf{q})^2(\mathbf{k} - \mathbf{k}')^2}. \quad (3.49)$$

Thus if $\mathcal{F}^{(8)}(\omega, \mathbf{k}, \mathbf{q})$ is independent of $\mathbf{k}$ we have

$$(\omega - \epsilon_G(-\mathbf{k}^2) - \epsilon_G(-(\mathbf{k} - \mathbf{q})^2))\mathcal{F}^{(8)}(\omega, .., \mathbf{q}) = \frac{\pi}{2} \frac{\alpha_s N \mathbf{q}^2}{4\pi^2}$$

$$+ (\epsilon_G(-\mathbf{q}^2) - \epsilon_G(-\mathbf{k}^2) - \epsilon_G(-(\mathbf{k} - \mathbf{q})^2))\mathcal{F}^{(8)}(\omega, .., \mathbf{q}). \quad (3.50)$$

The terms with factors of $\epsilon_G(-\mathbf{k}^2)$ and $\epsilon_G(-(\mathbf{k} - \mathbf{q})^2)$ cancel out exactly. It is worth emphasizing that this remarkable cancellation only works in the octet exchange channel. It depends crucially on the fact that the colour factor from the addition of an extra rung is $N/2$. It is the generalization of the seemingly miraculous cancellation of those terms corresponding to Figs. 3.5 and 3.8 which spoiled the exponentiation up to order α_s^2.

The solution to Eq.(3.50) is simply

$$\mathcal{F}^{(8)}(\omega, .., \mathbf{q}) = \frac{\pi}{2} \frac{\alpha_s N \mathbf{q}^2}{4\pi^2} \frac{1}{(\omega - \epsilon_G(-\mathbf{q}^2))}, \quad (3.51)$$

so that the imaginary part of the amplitude (inserting into Eq.(3.44) and recalling that $t = -\mathbf{q}^2$) is

$$\Im m A^{(8)}(s, t) = -\frac{\pi}{2} \epsilon_G(t) \left(\frac{s}{\mathbf{k}^2} \right)^{\epsilon_G(t)} \mathcal{A}_0^{(8)}. \quad (3.52)$$

The analytic function of which this is the imaginary part is

$$= \frac{1}{2} \left(\frac{-s}{\mathbf{k}^2} \right)^{\epsilon_G(t)} \mathcal{A}_0^{(8)}.$$

Since $\mathcal{A}_0^{(8)}$ is proportional to s we have an s-dependence of

$$-(-s)^{1+\epsilon_G(t)}.$$

Adding the contribution from the u-channel graphs and using $u \approx -s$ we obtain a total expression for the octet exchange amplitude from summing the leading $\ln s$ to all orders in perturbation theory given by

$$\mathcal{A}^{(8)} = 8\pi\alpha_s \frac{\mathbf{k}^2}{t} \tau^a \otimes \tau^a \delta_{\lambda_1\lambda_1'} \delta_{\lambda_2\lambda_2'} \left(\frac{s}{\mathbf{k}^2}\right)^{\alpha_G(t)} \frac{1 - e^{i\pi\alpha_G(t)}}{2}, \quad (3.53)$$

where

$$\alpha_G(t) = 1 + \epsilon_G(t).$$

This is a Regge trajectory of odd signature and we have justified the *ansatz* made in Eq.(3.39) for the 'reggeized' gluon propagator.

Although $\alpha_G(t)$ is infra-red divergent, if we regularize using dimensional regularization, i.e. if we perform the integration over transverse momenta in $2 + \epsilon$ dimensions, then we have[†]

$$\alpha_G(-\mathbf{q}^2) = 1 - \frac{N\alpha_s\mathbf{q}^2}{(2\pi)^{2+\epsilon}} \int \frac{d^{2+\epsilon}\mathbf{k}}{\mathbf{k}^2(\mathbf{k}-\mathbf{q})^2} = 1 - \frac{N\alpha_s}{4\pi} \frac{2(\mathbf{q}^2)^{\epsilon/2}}{\epsilon},$$

such that $\alpha_G(0) = 1$ and we find that the massless, spin-1 gluon does indeed lie on the trajectory. This has been shown by Fadin, Kuraev & Lipatov (1976), Frankfurt & Sherman (1976), Tyburski (1976), and Cheng & Lo (1976) to be true also in the case where the gauge group is broken spontaneously so that the 'gluon' acquires a mass, M, and it turns out that $\alpha_G(M^2) = 1$. In this case graphs involving Higgs bosons (which do not occur in the treatment described in this chapter) play a crucial role.

We have now done most of the hard work. In the next chapter we shall be using these reggeized gluons to construct the perturbative Pomeron.

3.8 Summary

• The first few terms in the perturbative expansion for the amplitude involving spin-1, colour octet exchange suggest that the gluon reggeizes, i.e. its propagator is given by Eq.(3.1) with $\epsilon_G(t)$ given by Eq.(3.9). After regularization of the infra-red divergence

---

[†] We have absorbed $\ln 4\pi$ and the Euler constant γ_E into $1/\epsilon$.

we find $\alpha_G(0)\,(= 1 + \epsilon_G(0)) = 1$ so that the gluon lies on this trajectory.

• The two-quark to two-quark plus n-gluon amplitude at the tree level, in the kinematic regime which leads to leading $\ln s$ in the octet exchange amplitude, is given by the left half of uncrossed ladder diagrams with effective vertices, $\Gamma^\sigma_{\mu\nu}$, given by Eq.(3.11) coupling the rungs of the ladder and the vertical gluon lines.

• Loop corrections in leading $\ln s$ approximation are introduced by replacing the propagators for the vertical gluon lines of the ladder by reggeized gluons.

• An integral equation for the Mellin transform of the imaginary part of the octet exchange amplitude can be obtained using a dispersion relation involving these ladders.

• The integral equation has a solution which consists of a simple pole at $\omega = \epsilon_G(t)$, thereby justifying the proposition that the gluon reggeizes.

4

The QCD Pomeron

Following the success of the reggeization of various different elementary particles it was hoped that a particle could be identified with the quantum numbers of the Pomeron which would reggeize to give the Pomeron trajectory.

Unfortunately this turned out not to be possible. In particular, in QCD all the elementary particles carry colour so there is no basic QCD constituent with the quantum numbers of the Pomeron. In QCD the lowest order Feynman diagram that can simulate the exchange of a Pomeron is a two-gluon exchange diagram. This led Low (1975) to use two-gluon exchange as a model for the bare Pomeron. He made numerical estimates of the amplitude for the exchange of two gluons between two hadrons using the then fashionable bag model of hadrons. This was then developed by Nussinov (1975, 1976), who considered contributions from more than two exchanged gluons as well as uncrossed ladder corrections to the two-gluon exchange amplitude.

We have already implicitly used the Low–Nussinov model in Chapter 2 to construct the Pomeron in the scalar theory model considered in that chapter. Combining this with our experience in deriving the reggeized gluon we can see what the picture of the Pomeron is in leading logarithm perturbative QCD.

The imaginary part of the amplitude for Pomeron exchange is given in terms of the multi-Regge exchange amplitude for two incoming particles (quarks for simplicity) to scatter into two quarks plus n gluons. The difference between this imaginary part and the imaginary part of the reggeized gluon lies solely in the colour factor. In the Pomeron case a singlet of colour is exchanged in the t-channel.

At the tree level this amplitude is just the left hand side of a ladder graph with triple gluon vertices replaced by the effective vertices, Γ, discussed in the preceding chapter. Loop effects are taken into account by replacing the vertical gluon lines of the ladder by reggeized gluons. In summary, the QCD Pomeron consists of a ladder whose vertical lines are reggeized gluons, with effective vertices, Γ, which couple the reggeized gluons and the rungs of the ladder, and with no colour carried up the ladder.

The Pomeron that is obtained has even charge conjugation, which means that it has the same coupling to quarks and to antiquarks. A trajectory with similar quantum numbers as the Pomeron but which has odd charge conjugation has been proposed by Bouquet *et al.* (1975) and Joynson *et al.* (1975) and is called the **odderon**. The lowest order diagram for odderon exchange is the exchange of three gluons in a colour singlet state. We shall not be considering the odderon in this book.

The famous Balitsky, Fadin, Kuraev, Lipatov (BFKL) equation is the integral equation which determines the behaviour of the Pomeron described above, in perturbative QCD. Several independent paths have led directly or indirectly to this integral equation. The method that we shall follow in this chapter is that of Fadin, Kuraev & Lipatov (1976) and Balitsky & Lipatov (1978).

4.1 First three orders of perturbation theory

Our task is to calculate, to leading $\ln s$, the amplitude for quark–quark elastic scattering, i.e. incoming quarks with momentum p_1, p_2 and helicity λ_1, λ_2 scatter into a final state of quarks with momentum $(p_1 - q), (p_2 + q)$ and helicity $\lambda_{1'}, \lambda_{2'}$, via the exchange of a colour singlet. Most of the work has already been done in the preceding chapter, when we considered the case of colour octet exchange in order to obtain the reggeized gluon. The difference arises in the colour factors. This chapter is therefore shorter and much less painful than the last!

Since we are interested in colour singlet exchange there is no contribution from the tree diagram Fig. 3.2. The lowest order which gives a non-trivial contribution is the one loop contribution shown in Fig. 3.4. In this case the colour factor is different from the case of the reggeized gluon. We project out the singlet contribution so the colour factor of both graphs in Fig. 3.4 is

$$G_0^{(1)} = \frac{1}{N^2} \text{Tr}(\tau_a \tau_b) \text{Tr}(\tau_a \tau_b) = \frac{N^2 - 1}{4N^2}. \tag{4.1}$$

Thus as can be seen from Eqs. (3.6) and (3.7) in the limit $|t| \ll s$ where $u \approx -s$, the real part of the amplitude cancels and we are left with a purely imaginary part which can be read off from Eq.(3.5):

$$A_1^{(1)} = 4i\alpha_s^2 s \delta_{\lambda_1' \lambda_1} \delta_{\lambda_2' \lambda_2} G_0^{(1)} \int \frac{d^2\mathbf{k}}{\mathbf{k}^2(\mathbf{k} - \mathbf{q})^2}. \tag{4.2}$$

Since the amplitude begins in order α_s^2 (with no $\ln s$ factor), i.e. order by order in perturbation theory the Pomeron exchange amplitude is suppressed by a power of $\ln s$ relative to the amplitude for the reggeized gluon exchange, it follows that the Pomeron has even signature. What we mean is that the amplitude for Pomeron exchange contains a signature factor:

$$\frac{1}{2}\left(1 + e^{i\pi\alpha_P(t)}\right),$$

where the Pomeron trajectory, $\alpha_P(t) = 1 + \mathcal{O}(\alpha_s)$. Expanding the above signature factor as a power series in α_s we see that the leading non-trivial order is imaginary and $\mathcal{O}(\alpha_s)$.

In the next order of perturbation theory the amplitude for colour singlet exchange has two components (as was the case for the reggeized gluon). The first component is given by (see Eq.(3.17))

$$A_{2a}^{(1)} = i\frac{-g_{\sigma\tau}}{2} \int d(P.S.^3) A_{2\to3}^{(8)\sigma}(k_1, k_2) A_{2\to3}^{\dagger(8)\tau}(k_1 - q, k_2 - q), \tag{4.3}$$

but in this case the colour factor is given by

$$\frac{1}{N} \text{Tr}(\tau_a \tau_b) \text{Tr}(\tau_c \tau_d) f_{ace} f_{bde} = N G_0^{(1)}.$$

Thus from Eq.(3.22) we see that this gives us a contribution to the colour singlet amplitude:

$$A_{2a}^{(1)} = -i\frac{2N\alpha_s^3}{\pi^2} \delta_{\lambda_1' \lambda_1} \delta_{\lambda_2' \lambda_2} G_0^{(1)} s \ln(s/\mathbf{k}^2) \int d^2\mathbf{k}_1 d^2\mathbf{k}_2$$

$$\times \left[\frac{\mathbf{q}^2}{\mathbf{k}_1^2 \mathbf{k}_2^2 (\mathbf{k}_1 - \mathbf{q})^2 (\mathbf{k}_2 - \mathbf{q})^2} - \frac{1}{\mathbf{k}_1^2 (\mathbf{k}_1 - \mathbf{k}_2)^2 (\mathbf{k}_2 - \mathbf{q})^2} \right.$$

$$\left. - \frac{1}{\mathbf{k}_2^2 (\mathbf{k}_1 - \mathbf{q})^2 (\mathbf{k}_1 - \mathbf{k}_2)^2} \right]. \tag{4.4}$$

The other component comes from the diagrams of Fig. 3.8. In the colour singlet case the colour factor is given by

$$\frac{1}{N^2}\text{Tr}(\tau_a\tau_b\tau_c)\text{Tr}(\tau_a\tau_b\tau_c) = \frac{N}{2}G_0^{(1)}.$$

So from Eq.(3.24) and this colour factor we see that the contribution from these graphs is

$$
\begin{aligned}
\mathcal{A}_{2b}^{(1)} &= -i\frac{N\alpha_s^3}{\pi^2}\delta_{\lambda_1'\lambda_1}\delta_{\lambda_2'\lambda_2}G_0^{(1)}s\ln\left(s/\mathbf{k}^2\right) \\
&\quad \times \int d^2\mathbf{k_1}d^2\mathbf{k_2}\frac{1}{\mathbf{k_1^2}(\mathbf{k_1}-\mathbf{k_2})^2(\mathbf{k_2}-\mathbf{q})^2},
\end{aligned}
\tag{4.5}
$$

with a similar contribution coming from the graphs with one gluon on the left of the cut.

We note that in this case we do *not* get a cancellation of the non-factorizing part and so we do *not* obtain an expression at the two-loop level which is proportional to the one loop amplitude. This is due to the different colour factors.

The total expression to order α_s^3 is therefore

$$
\begin{aligned}
\Im m\mathcal{A}_2^{(1)} &= -\frac{2N\alpha_s^3}{\pi^2}s\,\delta_{\lambda_1'\lambda_1}\delta_{\lambda_2'\lambda_2}G_0^{(1)}\ln\left(s/\mathbf{k}^2\right)\int d^2\mathbf{k_1}d^2\mathbf{k_2} \\
&\quad \times \left[\frac{\mathbf{q}^2}{\mathbf{k_1^2}\mathbf{k_2^2}(\mathbf{k_1}-\mathbf{q})^2(\mathbf{k_2}-\mathbf{q})^2}\right. \\
&\quad \left. -\frac{1}{2}\frac{1}{\mathbf{k_1^2}(\mathbf{k_1}-\mathbf{k_2})^2(\mathbf{k_2}-\mathbf{q})^2} -\frac{1}{2}\frac{1}{\mathbf{k_2^2}(\mathbf{k_1}-\mathbf{k_2})^2(\mathbf{k_1}-\mathbf{q})^2}\right].
\end{aligned}
\tag{4.6}
$$

It will once again prove to be convenient to work in terms of the Mellin transform of the amplitude and to this end we define a function $f(\omega,\mathbf{k_1},\mathbf{k_2},\mathbf{q})$ by

$$
\begin{aligned}
\int_1^\infty d\left(\frac{s}{\mathbf{k}^2}\right)\left(\frac{s}{\mathbf{k}^2}\right)^{-\omega-1}\frac{\mathcal{A}^{(1)}(s,t)}{s} &= 4i\alpha_s^2\delta_{\lambda_1'\lambda_1}\delta_{\lambda_2'\lambda_2}G_0^{(1)} \\
&\quad \times \int\frac{d^2\mathbf{k_1}d^2\mathbf{k_2}}{\mathbf{k_2^2}(\mathbf{k_1}-\mathbf{q})^2}f(\omega,\mathbf{k_1},\mathbf{k_2},\mathbf{q})
\end{aligned}
\tag{4.7}
$$

(note that the amplitude has been divided by s before the Mellin transform has been taken so that the leading order term has a Mellin transform proportional to $1/\omega$).

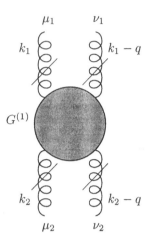

Fig. 4.1. Four-gluon Green function.

$f(\omega, \mathbf{k_1}, \mathbf{k_2}, \mathbf{q})$ is related to a Green function with four off-shell external gluons (see Fig. 4.1) by

$$\frac{f(\omega, \mathbf{k_1}, \mathbf{k_2}, \mathbf{q})}{\mathbf{k_2^2}(\mathbf{k_1} - \mathbf{q})^2} = \frac{s}{(2\pi)^4} \int d\rho_1 d\lambda_2 \frac{4 p_1^{\mu_1} p_1^{\nu_1} p_2^{\mu_2} p_2^{\nu_2}}{s^2}$$

$$\times \, G^{(1)}_{\mu_1 \mu_2 \nu_1 \nu_2}(\omega, k_1, k_2, q), \qquad (4.8)$$

where $G^{(1)}_{\mu_1 \mu_2 \nu_1 \nu_2}(\omega, k_1, k_2, q)$ is the Mellin transform of the Green function for four external gluons with momenta $k_1, k_2, k_1 - q, k_2 - q$ with gluons 1 and 3 (2 and 4) being in a colour singlet state. These momenta can be expressed in terms of Sudakov variables ρ_1, λ_2 and their transverse components as

$$k_1^\mu = \rho_1 p_1^\mu - \frac{\mathbf{k_1^2}}{s} p_2^\mu + k_{1\perp}^\mu,$$

$$k_2^\mu = \frac{\mathbf{k_2^2}}{s} p_1^\mu + \lambda_2 p_2^\mu + k_{2\perp}^\mu,$$

$$q^\mu = \frac{\mathbf{q^2}}{s} p_1^\mu - \frac{\mathbf{q^2}}{s} p_2^\mu + q_\perp^\mu,$$

with s $(= 2 p_1 \cdot p_2) \gg |\mathbf{q^2}|$. The integral over ρ_1, λ_2 is dominated by the region

$$\frac{\mathbf{k^2}}{s} \ll \rho_1, |\lambda_2| \ll 1.$$

The definition of $f(\omega, \mathbf{k_1}, \mathbf{k_2}, \mathbf{q})$ is somewhat different from the definition of $\mathcal{F}^{(8)}(\omega, \mathbf{k}, \mathbf{q})$ used in the case of the reggeized gluon, but has the advantage of being symmetric in $\mathbf{k_1}$ and $\mathbf{k_2}$ and can be viewed as a Green function.

Thus in leading (non-trivial) order of perturbation theory we have

$$f_1(\omega, \mathbf{k_1}, \mathbf{k_2}, \mathbf{q}) = \frac{1}{\omega} \delta^2(\mathbf{k_1} - \mathbf{k_2}) \qquad (4.9)$$

and in the next order

$$f_2(\omega, \mathbf{k_1}, \mathbf{k_2}, \mathbf{q}) = -\frac{\bar{\alpha}_s}{2\pi} \frac{1}{\omega^2}$$

$$\times \left[\frac{\mathbf{q}^2}{\mathbf{k_1^2}(\mathbf{k_2} - \mathbf{q})^2} - \frac{1}{2} \frac{1}{(\mathbf{k_1} - \mathbf{k_2})^2} \left(1 + \frac{\mathbf{k_2^2}(\mathbf{k_1} - \mathbf{q})^2}{\mathbf{k_1^2}(\mathbf{k_2} - \mathbf{q})^2} \right) \right]. \quad (4.10)$$

For convenience we choose to define the commonly recurring factor

$$\bar{\alpha}_s = \frac{N\alpha_s}{\pi}. \qquad (4.11)$$

In the next section we discuss how to calculate the leading (ω-plane) singularity of $f(\omega, \mathbf{k_1}, \mathbf{k_2}, \mathbf{q})$ which determines the leading logarithm contribution to the amplitude $\mathcal{A}^{(1)}(s, t)$.

4.2 The BFKL equation

As discussed above, the leading logarithm contribution to the colour singlet exchange amplitude is given by the infinite sum of ladders in which the vertical lines are reggeized gluons, and the couplings to the horizontal rungs are given by the effective vertices, Γ, of Eq. (3.11), but with a colour factor that projects colour singlet exchange.

One may ask why we only allow reggeized gluons in the vertical lines. Is it not possible that some of the vertical lines are Pomerons themselves, so that we have a similar bootstrap to that which we found in the case of the reggeized gluon? The answer to this goes back to the statement that the Pomeron starts in perturbation theory at one order in α_s higher than the reggeized gluon. The replacement, therefore, of any one of the reggeized gluons in the vertical lines by a Pomeron gives a contribution in any order of perturbation theory which is suppressed by a factor of $\ln s$ and is thus neglected in the leading logarithm approximation.

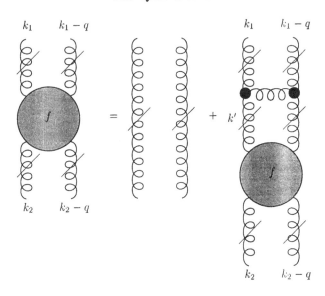

Fig. 4.2. Integral equation for f.

The quantity $f(\omega, \mathbf{k_1}, \mathbf{k_2}, \mathbf{q})$ is therefore given by an integral equation analogous to Eq.(3.46) for the octet quantity $\mathcal{F}^{(8)}(\omega, \mathbf{k}, \mathbf{q})$. However, the addition of an extra rung introduces a colour factor of

$$\frac{\delta_{cd} f_{cae} f_{dbe}}{\delta_{ab}} = N$$

rather than $N/2$ as was the case for colour octet exchange. Once again the Born term is just the exchange of two reggeized gluons and the integral equation equivalent of Eq.(3.46) is shown in Fig. 4.2 and reads

$$(\omega - \epsilon_G(-\mathbf{k_1^2}) - \epsilon_G(-(\mathbf{k_1} - \mathbf{q})^2)) f(\omega, \mathbf{k_1}, \mathbf{k_2}, \mathbf{q})$$

$$= \delta^2(\mathbf{k_1} - \mathbf{k_2}) - \frac{\bar{\alpha}_s}{2\pi} \int d^2 k' \left[\frac{\mathbf{q}^2}{(\mathbf{k'} - \mathbf{q})^2 \mathbf{k_1^2}} \right.$$

$$\left. - \frac{1}{(\mathbf{k'} - \mathbf{k_1})^2} \left(1 + \frac{(\mathbf{k_1} - \mathbf{q})^2 \mathbf{k'^2}}{(\mathbf{k'} - \mathbf{q})^2 \mathbf{k_1^2}} \right) \right] f(\omega, \mathbf{k'}, \mathbf{k_2}, \mathbf{q}). \qquad (4.12)$$

In the special case of zero momentum transfer, i.e. $\mathbf{q} = 0$, we can simplify this equation to read,

$$[\omega - 2\epsilon_G(-\mathbf{k}_1^2)]f(\omega, \mathbf{k}_1, \mathbf{k}_2, 0) = \delta^2(\mathbf{k}_1 - \mathbf{k}_2)$$
$$+ \bar{\alpha}_s \int \frac{d^2\mathbf{k}'}{\pi} \frac{f(\omega, \mathbf{k}', \mathbf{k}_2, 0)}{(\mathbf{k}' - \mathbf{k}_1)^2}. \qquad (4.13)$$

Equation (4.12) has a remarkable property – *it is infra-red finite.* In order to see this we use Eqs.(3.48) and (3.49) and exploit the shift of integration variable $\mathbf{k}' \rightarrow (\mathbf{k}_1 - \mathbf{k}')$ which allows us to make the replacements

$$\frac{1}{\mathbf{k}'^2(\mathbf{k}_1 - \mathbf{k}')^2} \rightarrow \frac{2}{(\mathbf{k}_1 - \mathbf{k}')^2[\mathbf{k}'^2 + (\mathbf{k}_1 - \mathbf{k}')^2]} \qquad (4.14)$$

and

$$\frac{1}{(\mathbf{k}' - \mathbf{q})^2(\mathbf{k}_1 - \mathbf{k}')^2} = \frac{2}{(\mathbf{k}_1 - \mathbf{k}')^2[(\mathbf{k}' - \mathbf{q})^2 + (\mathbf{k}_1 - \mathbf{k}')^2]}. \qquad (4.15)$$

Equation (4.12) may then be rewritten:

$$\omega f(\omega, \mathbf{k}_1, \mathbf{k}_2, \mathbf{q}) = \delta^2(\mathbf{k}_1 - \mathbf{k}_2)$$
$$+ \frac{\bar{\alpha}_s}{2\pi} \int d^2\mathbf{k}' \left[\frac{-\mathbf{q}^2}{(\mathbf{k}' - \mathbf{q})^2\mathbf{k}_1^2} f(\omega, \mathbf{k}', \mathbf{k}_2, \mathbf{q}) \right.$$
$$+ \frac{1}{(\mathbf{k}' - \mathbf{k}_1)^2} \left(f(\omega, \mathbf{k}', \mathbf{k}_2, \mathbf{q}) - \frac{\mathbf{k}_1^2 f(\omega, \mathbf{k}_1, \mathbf{k}_2, \mathbf{q})}{\mathbf{k}'^2 + (\mathbf{k}_1 - \mathbf{k}')^2} \right) .$$
$$+ \frac{1}{(\mathbf{k}' - \mathbf{k}_1)^2} \left(\frac{(\mathbf{k}_1 - \mathbf{q})^2\mathbf{k}'^2 f(\omega, \mathbf{k}', \mathbf{k}_2, \mathbf{q})}{(\mathbf{k}' - \mathbf{q})^2\mathbf{k}_1^2} \right.$$
$$\left. \left. - \frac{(\mathbf{k}_1 - \mathbf{q})^2 f(\omega, \mathbf{k}_1, \mathbf{k}_2, \mathbf{q})}{(\mathbf{k}' - \mathbf{q})^2 + (\mathbf{k}_1 - \mathbf{k}')^2} \right) \right]. \qquad (4.16)$$

This is the BFKL equation.

The infra-red finiteness can now be seen by observing that the terms in parentheses multiplying the factor $1/(\mathbf{k}' - \mathbf{k}_1)^2$ vanish at $\mathbf{k}_1 = \mathbf{k}'$. It was in order to make this explicit that the manipulations Eqs.(4.14) and (4.15) were employed. This cancellation justifies our hitherto cavalier treatment of infra-red divergent integrals. This cancellation of infra-red divergences has also been demonstrated by Jaroszewicz (1980) using Ward identities and working in Coulomb gauge.

In fact, the cancellation of the infra-red divergences can be used to justify *a posteriori* the use of the strong ordering of the longitudinal components of momenta (i.e. the multi-Regge regime). We have established, in Chapter 2, that the leading logarithm contribution to the integration over longitudinal momenta requires the multi-Regge kinematics. This provides the leading logarithm contribution *provided there are no further logarithms generated by the integration over transverse momenta*. The infra-red finiteness of the BFKL equation means that no such extra logarithms can occur. It is important to appreciate that, in order to ensure the infra-red finiteness, we had to integrate over all intermediate states (of the cut amplitude). It has been pointed out by Marchesini (1995) that for some associated (i.e. not fully inclusive) distributions the infra-red finiteness is lost and consequently the multi-Regge kinematics no longer leads to the leading logarithm contribution. We shall return to this matter at the end of Chapter 6.

4.3 The solution for zero momentum transfer

To keep the mathematics simpler, we first consider the solution to Eq.(4.16) in the case where $\mathbf{q} = 0$ (i.e. we look at the intercept of the QCD Pomeron at $t = 0$). In this case, the BFKL equation becomes

$$\omega f(\omega, \mathbf{k_1}, \mathbf{k_2}, 0) = \delta^2(\mathbf{k_1} - \mathbf{k_2}) + \mathcal{K}_0 \bullet f(\omega, \mathbf{k_1}, \mathbf{k_2}, 0), \quad (4.17)$$

where

$$\mathcal{K}_0 \bullet f(\omega, \mathbf{k_1}, \mathbf{k_2}, 0) = \frac{\bar{\alpha}_s}{\pi} \int \frac{d^2\mathbf{k'}}{(\mathbf{k_1} - \mathbf{k'})^2}$$

$$\times \left[f(\omega, \mathbf{k'}, \mathbf{k_2}, 0) - \frac{\mathbf{k}_1^2}{[\mathbf{k'}^2 + (\mathbf{k_1} - \mathbf{k'})^2]} f(\omega, \mathbf{k_1}, \mathbf{k_2}, 0) \right]. \quad (4.18)$$

This is a Green function equation which is solved if we can find the complete set of eigenfunctions, $\phi_i(\mathbf{k})$ (with eigenvalues λ_i), of the integral operator (or **kernel**), $\mathcal{K}_0$, i.e.

$$\mathcal{K}_0 \bullet \phi_i(\mathbf{k}) = \lambda_i \phi_i(\mathbf{k}).$$

The eigenfunctions must obey the completeness relation

$$\sum_i \phi_i(\mathbf{k_1})\phi_i^*(\mathbf{k_2}) = \delta^2(\mathbf{k_1} - \mathbf{k_2}),$$

where the sum over the eigenfunction label, i, may involve an integral over a continuous variable. The solution to the Green function equation is then given by

$$f(\omega, \mathbf{k_1}, \mathbf{k_2}, 0) = \sum_i \frac{\phi_i(\mathbf{k_1})\phi_i^*(\mathbf{k_2})}{\omega - \lambda_i}. \qquad (4.19)$$

Since $\mathbf{k}$ is a vector in the two-dimensional transverse space we can write

$$\mathbf{k} = (k, \theta)$$
$$\mathbf{k}' = (k', \theta')$$

and

$$d^2\mathbf{k}' = \frac{1}{2}dk'^2 d\theta'.$$

By Fourier analysis $\phi_i(\mathbf{k})$ can be expanded in powers of $\exp(i\theta)$ with coefficients $\phi_i^n(k)$

$$\phi_i(\mathbf{k}) = \sum_{n=0}^{\infty} \phi_i^n(k) \frac{e^{in\theta}}{\sqrt{2\pi}}.$$

Inserting each of these components into $\mathcal{K}_0 \bullet \phi_i$ and performing the angular integral over θ' gives

$$\mathcal{K}_0 \bullet \phi_i^n(k) = \bar{\alpha}_s e^{in\theta} \int dk'^2 \left\{ \frac{1}{|k'^2 - k^2|} \right.$$
$$\times \left[\left(\frac{k'k}{\max(k^2, k'^2)} \right)^n \phi_i^n(k') - \frac{k^2}{\max(k^2, k'^2)} \phi_i^n(k) \right]$$
$$\left. + \frac{4\phi_i^n(k)}{\sqrt{4k'^4 + k^4}} \left[\frac{k'^2\theta(k^2 - k'^2)}{k^2 + \sqrt{4k'^4 + k^4}} - \theta(k'^2 - k^2)\frac{k^2}{k'^2} \right] \right\}. \qquad (4.20)$$

After the integral over k'^2 the last two terms cancel each other, in anticipation of which we rewrite the right hand side of Eq.(4.20) as

$$\bar{\alpha}_s e^{in\theta} \int_0^{k^2} \frac{dk'^2}{k^2 - k'^2} \left[\left(\frac{k'}{k} \right)^n \phi_i^n(k') - \phi_i^n(k) \right]$$
$$+ \bar{\alpha}_s e^{in\theta} \int_{k^2}^{\infty} \frac{dk'^2}{k'^2 - k^2} \left[\left(\frac{k}{k'} \right)^n \phi_i^n(k') - \frac{k^2}{k'^2} \phi_i^n(k) \right]. \qquad (4.21)$$

Now since there is no infra-red divergence (i.e. no need to introduce a dimensionful scale to regularize the integrals in Eq.(4.21)) and since the kernel is a dimensionless operator, it follows that

$\phi_i^n(k)$ behaves like a power of k^2. In order to have a set of eigen-functions that obey the completeness relation we need to restrict this power behaviour to the form

$$\phi_\nu(k) \sim (k^2)^{-1/2+i\nu}, \qquad (4.22)$$

where $-\infty < \nu < \infty$. Thus the complete eigenfunctions are

$$\phi_\nu^n(\mathbf{k}) = \frac{1}{\pi\sqrt{2}}(k^2)^{-1/2+i\nu}e^{in\theta}. \qquad (4.23)$$

These are normalized so as to satisfy

$$\int d^2\mathbf{k}\,\phi_\nu^n(\mathbf{k})\phi_{\nu'}^{n'*}(\mathbf{k}) = \delta(\nu - \nu')\delta(n - n'). \qquad (4.24)$$

To find the eigenvalues we insert the function $\phi_\nu^n(\mathbf{k})$ into Eq.(4.21) and obtain

$$e^{in\theta}\phi_\nu^n(\mathbf{k})\bar{\alpha}_s\left[\int_0^1 dz\frac{(z^{(n-1)/2+i\nu} - 1)}{(1 - z)} + \int_0^1 dw\frac{(w^{(n-1)/2-i\nu} - 1)}{(1 - w)}\right],$$

where

$$z = \frac{k'^2}{k^2} \quad \text{and} \quad w = \frac{k^2}{k'^2}.$$

Hence the eigenvalue, $\omega_n(\nu)$, is

$$\omega_n(\nu) = \bar{\alpha}_s\chi_n(\nu), \qquad (4.25)$$

where

$$\chi_n(\nu) = 2\int_0^1 dz\frac{z^{(n-1)/2}\cos(\nu\ln z) - 1}{(1 - z)}. \qquad (4.26)$$

This is a standard integral which is given in terms of the digamma function, ψ (the logarithmic derivative of the Γ function), i.e.

$$\chi_n(\nu) = 2\left(-\gamma_E - \Re e\left[\psi\left((n + 1)/2 + i\nu\right)\right]\right), \qquad (4.27)$$

where $\gamma_E \approx 0.577$ is Euler's constant.

Thus the solution for $f(\omega, \mathbf{k_1}, \mathbf{k_2}, 0)$ is

$$f(\omega, \mathbf{k_1}, \mathbf{k_2}, 0) = \sum_{n=0}^{\infty}\int_{-\infty}^{\infty} d\nu\left(\frac{k_1^2}{k_2^2}\right)^{i\nu}\frac{e^{in(\theta_1 - \theta_2)}}{2\pi^2 k_1 k_2}\frac{1}{(\omega - \bar{\alpha}_s\chi_n(\nu))}. \qquad (4.28)$$

Our first observation is that since ν is a continuous variable we do not obtain an isolated pole in the Mellin transform which we can associate with the intercept of the Pomeron. Leading logarithm perturbation theory gives us a cut rather than a pole. We shall return to this matter later.

We are interested in the leading $\ln s$ behaviour which means the singularity with the largest real part in the ω-plane. This allows us to make a number of simplifications. Since the function $\chi_n(\nu)$ decreases with increasing n, we are at liberty to restrict the sum over n in Eq.(4.28) to the case where $n = 0$. Furthermore, $\chi_0(\nu)$ decreases with increasing $|\nu|$ so we can expand $\chi_0(\nu)$ as a power series in ν and keep only the first two terms. We obtain

$$\chi_0(\nu) = 4\ln 2 - 14\zeta(3)\nu^2 + \cdots, \qquad (4.29)$$

with $\zeta(3) = \sum_r (1/r^3) \approx 1.202$. In this approximation

$$f(\omega, \mathbf{k_1}, \mathbf{k_2}, 0) \approx \frac{1}{\pi k_1 k_2} \int_{-\infty}^{\infty} \frac{d\nu}{2\pi} \left(\frac{k_1^2}{k_2^2}\right)^{i\nu} \frac{1}{(\omega - \omega_0 + a^2\nu^2)}, \qquad (4.30)$$

with

$$\omega_0 = 4\bar{\alpha}_s \ln 2 \qquad (4.31)$$

being the position of the leading singularity (the branch point of the cut) and

$$a^2 = 14\bar{\alpha}_s \zeta(3). \qquad (4.32)$$

We can perform the integration over ν (using contour integration) and obtain

$$f(\omega, \mathbf{k_1}, \mathbf{k_2}, 0) \approx \frac{1}{2\pi a k_1 k_2} \left(\frac{k_1 k_2}{\max(k_1^2, k_2^2)}\right)^{\sqrt{\omega - \omega_0}/a} \frac{1}{\sqrt{\omega - \omega_0}}. \qquad (4.33)$$

Moreover, one can invert the Mellin transform to expose the s-dependence of the colour singlet amplitude. This is most easily effected by inverting Eq.(4.30) before integrating over ν. We find

$$F(s, \mathbf{k_1}, \mathbf{k_2}, 0) \approx \frac{1}{\sqrt{\mathbf{k_1^2 k_2^2}}} \left(\frac{s}{\mathbf{k}^2}\right)^{\omega_0} \frac{1}{\sqrt{\pi \ln(s/\mathbf{k}^2)}}$$
$$\times \frac{1}{2\pi a} \exp\left(-\frac{\ln^2(\mathbf{k_1^2}/\mathbf{k_2^2})}{4a^2 \ln(s/\mathbf{k}^2)}\right). \qquad (4.34)$$

This is the inverse transform of Eq.(4.33), i.e. the full amplitude for quark–quark forward elastic scattering is simply (see Eq.(4.7))

$$\frac{A^{(1)}(s, 0)}{s} = 4i\alpha_s^2 \delta_{\lambda_1'\lambda_1} \delta_{\lambda_2'\lambda_2} G_0^{(1)} \int \frac{d^2\mathbf{k_1}}{\mathbf{k_1^2}} \frac{d^2\mathbf{k_2}}{\mathbf{k_2^2}} F(s, \mathbf{k_1}, \mathbf{k_2}, 0). \qquad (4.35)$$

Note the factor of $1/\sqrt{\ln s}$. It arises from the fact that the Mellin transform contains an ω-plane cut rather than a simple pole.

The effect of including higher order terms in the expansion of $\chi_0(\nu)$ is to add terms which are suppressed by powers of $\ln s$, and so we are formally justified in neglecting them.

4.4 Impact factors

Before we move on to the somewhat more complicated problem of solving the BFKL equation for non-zero momentum transfer, we digress a little to discuss the matter of impact factors.

So far we have considered only quark–quark elastic scattering, where the external quarks are on shell. In practice, this is not what actually happens: the Pomeron couples to a hadron inside which the partons are slightly off-shell. Indeed, in the case of quark–quark scattering, despite the fact that $f(\omega, \mathbf{k_1}, \mathbf{k_2}, \mathbf{q})$ does not contain any infra-red singularities, the amplitude nevertheless diverges owing to the remaining integrals over $\mathbf{k_1}$ and $\mathbf{k_2}$ which develop infra-red singularities when $\mathbf{k_1}$ or $\mathbf{k_2}$ go to zero (or when $(\mathbf{k_1} - \mathbf{q})$ or $(\mathbf{k_2} - \mathbf{q})$ go to zero). These infra-red divergences are regulated by the slight off-shellness of the quarks (or gluons) to which the QCD Pomeron couples inside the hadron.

This leads us to introduce the quantity Φ, which is called the **impact factor** and accounts for the coupling of the Pomeron to the hadrons. We will consider here the case of elastic hadron–hadron scattering.

For elastic scattering of a hadron with initial momentum p_1 and a hadron with initial momentum p_2 (and final momenta $p_1 - q$ and $p_2 + q$ respectively), the Mellin transform of the scattering amplitude is given by (see Fig. 4.3)[†]

$$\mathcal{A}^{(1)}(\omega, t) = \frac{\mathcal{G}}{(2\pi)^4} \int d^2\mathbf{k_1} d^2\mathbf{k_2} \frac{\Phi_1(\mathbf{k_1}, \mathbf{q})\Phi_2(\mathbf{k_2}, \mathbf{q})}{\mathbf{k_2^2}(\mathbf{k_1} - \mathbf{q})^2} f(\omega, \mathbf{k_1}, \mathbf{k_2}, \mathbf{q}),$$

(4.36)

where Φ_1 and Φ_2 are the impact factors associated with the two scattering hadrons. The factor, $\mathcal{G}$, is the colour factor for the process.

[†] Note that $\mathcal{A}^{(1)}(\omega, t)$ is the Mellin transform of $\Im m \mathcal{A}^{(1)}(s, t)/s$ as defined in Eq. (4.7).

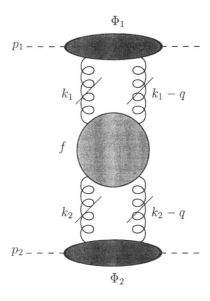

Fig. 4.3. Pomeron coupling to hadrons.

With these definitions, for the process of quark–quark elastic scattering the colour factor is

$$\mathcal{G} = G_0^{(1)}$$

and the impact factor is

$$\Phi_q = 2\pi g^2 \delta_{\lambda\lambda'}. \tag{4.37}$$

In order to calculate the impact factors we would need to know a great deal of detail about the wavefunction of the partons inside the hadron. Since this information is generally not available, models have to be used to calculate these impact factors. We will take a very simple model, namely, we consider meson–meson scattering and assume that the quarks are scalar particles and that the meson couples to quarks via a point-like coupling with dimensionful coupling constant, h. This last simplification just means that we do not have to worry about taking traces of Dirac matrices, and simplifies the expression that we obtain, but it does not qualitatively alter the result (the more physical case of spin-$\frac{1}{2}$ quarks coupling to vector photons is examined in the appendix to Chapter 6). In order to regulate the infra-red divergences we will

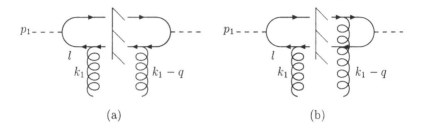

Fig. 4.4. Graphs for calculating impact factors.

have to introduce a quark mass, m, but we shall assume that the mesons remain massless.

The diagrams contributing to Φ_1 are shown in Fig. 4.4. We only need to calculate the leading order contribution since all higher order corrections are included in the quantity $f(\omega, \mathbf{k_1}, \mathbf{k_2}, \mathbf{q})$. Since we are using dispersive techniques to calculate the imaginary part of the amplitude (recall that the Pomeron amplitude is purely imaginary in the leading logarithm approximation) we consider the cut diagrams shown.

The momenta in the diagrams of Fig. 4.4 are labelled in such a way that the cut lines have momenta $l - k_1$ and $p_1 - l$ and so the two-body phase space may be written

$$d(P.S.^2) \;=\; \int \frac{d^4 k_1}{(2\pi)^3} \frac{d^4 l}{(2\pi)^3} \delta((k_1 - l)^2 - m^2)\delta((p_1 - l)^2 - m^2).$$

The diagram Fig. 4.4(a) leads to the amplitude,

$$A_{(a)}^{\mu\nu} = 4\pi\alpha_s h^2 \frac{(2l - k_1)^\mu (2l - k_1 - q)^\nu}{(l^2 - m^2)((l - q)^2 - m^2)}, \tag{4.38}$$

whereas the diagram of Fig. 4.4(b) gives

$$A_{(b)}^{\mu\nu} = 4\pi\alpha_s h^2 \frac{(2l - k_1)^\mu (2l - 2p_1 - k_1 + q)^\nu}{(l^2 - m^2)((l - p_1 - k_1 + q)^2 - m^2)}. \tag{4.39}$$

We have suppressed the colour matrices since they lead to the final colour factor of $\mathcal{G} = N^2 G_0^{(1)}$ (the factor N^2 arises because we no longer average over the incident quark colours).

As usual, we introduce Sudakov variables for l and k_1, i.e.

$$l^\mu = \rho p_1^\mu + \lambda p_2^\mu + l_\perp^\mu$$
$$k_1^\mu = \rho_1 p_1^\mu + \lambda_1 p_2^\mu + k_{1\perp}^\mu.$$

Recall that we are working in the eikonal approximation for which $\rho_1 \ll 1$ and so we are only interested in terms which are proportional to $p_1^\mu p_1^\nu$ in the numerator. The numerators then simplify to $4\rho^2 p_1^\mu p_1^\nu$ for Fig. 4.4(a) and $-4\rho(1-\rho)p_1^\mu p_1^\nu$ for Fig. 4.4(b). The limits on ρ are actually $\rho_1 < \rho < 1$, but since $\rho_1 \ll 1$ we can neglect ρ_1 compared with ρ up to corrections of order $\mathbf{k}_1^2/s$, $\mathbf{q}^2/s$.

In terms of the Sudakov variables the phase-space integral then becomes

$$d(P.S.^2) = \frac{1}{(2\pi)^6} \frac{1}{4} \int d\rho\, d\rho_1\, d\lambda\, d\lambda_1\, d^2\mathbf{l}\, d^2\mathbf{k_1}$$

$$\times\ \delta((1-\rho)\lambda + \mathbf{l}^2/s + m^2/s)\delta(\rho(\lambda-\lambda_1) - (1-\mathbf{k_1})^2/s + m^2/s).$$

After using the delta functions to fix λ and λ_1, and the on-shell condition for the final state mesons, $2p_1.q = -\mathbf{q}^2$, we find

$$A_{(a)}^{\mu\nu} = \frac{4\pi\alpha_s h^2\ 4\rho^2(1-\rho)^2\ p_1^\mu p_1^\nu}{(\mathbf{l}^2+m^2)(\mathbf{l}^2+\mathbf{q}^2(1-\rho)^2 - 2\mathbf{q}\cdot\mathbf{l}(1-\rho) + m^2)} \quad (4.40)$$

and

$$A_{(b)}^{\mu\nu} = -\frac{4\pi\alpha_s h^2\ 4\rho^2(1-\rho)^2\ p_1^\mu p_1^\nu}{(\mathbf{l}^2+m^2)((1-\mathbf{k_1})^2 + 2\rho\mathbf{q}\cdot(1-\mathbf{k_1}) + \mathbf{q}^2\rho^2 + m^2)}. \quad (4.41)$$

We now turn to the integral over the phase space. The integration over the transverse momentum $\mathbf{l}$ is most easily effected by introducing a Feynman parameter, τ, to combine the denominators, i.e. we use

$$\frac{1}{AB} = \int_0^1 d\tau \frac{1}{[A + \tau(B-A)]^2}. \quad (4.42)$$

Thus we find that,

$$2\int d(P.S.^2)(A_{(a)}^{\mu\nu} + A_{(b)}^{\mu\nu})$$

$$= \frac{\alpha_s h^2}{(2\pi)^4} 2p_1^\mu p_1^\nu \int d\rho\, d\rho_1\, d^2\mathbf{k_1}\, d\tau\, \rho(1-\rho)$$

$$\times \left[\frac{1}{(\mathbf{q}^2\rho^2\tau(1-\tau)+m^2)} - \frac{1}{[(\mathbf{k_1}-\rho\mathbf{q})^2\tau(1-\tau)+m^2]}\right] \quad (4.43)$$

and we have multiplied by 2 to include the related graphs in which the gluons couple to the opposite quark lines from those shown in Fig. 4.4. Note that the contribution from Fig. 4.4(a) is minus the contribution from Fig. 4.4(b) with $\mathbf{k_1}$ set to zero. The minus sign can be understood from the fact that in Fig. 4.4(a) the gluon on the right of the cut couples to the antiquark whilst in Fig. 4.4(b) it couples to a quark. This guarantees the vanishing of the impact factor Φ_1 when $\mathbf{k_1} = 0$, thereby regularizing the remaining infra-red divergence arising from the integration over $\mathbf{k_1}$. The infra-red finiteness of the colour singlet exchange amplitude can be interpreted as the cancellation between divergences arising from soft virtual gluon corrections (which are associated with the reggeization of the gluons in the vertical lines of the ladder as explained in the preceding chapter) and gluon bremsstrahlung (associated with adding more rungs to the ladder). The cancellation occurs for colour singlet amplitudes in accordance with the Kinoshita (1962), Lee, Nauenberg (1964) theorem. This theorem was originally derived for the case of QED. For non-abelian gauge theories it also works when applied to processes with colour singlet external states but *not* for colour non-singlet exchange amplitudes. That is why the Pomeron exchange amplitude is infra-red finite but the reggeized gluon is not.

There is a corresponding factor to that of Eq.(4.43) arising from the lower meson–Pomeron vertex (obtained by making the replacement $\mathbf{k_1} \to \mathbf{k_2}$ and integrating over λ_2 rather than ρ_1). Each of these factors must then be contracted into the four-gluon Green function. From Eq.(4.8), which relates $f(\omega, \mathbf{k_1}, \mathbf{k_2}, \mathbf{q})$ to the Green function, and Eq.(4.36), which defines the impact factors in terms of $f(\omega, \mathbf{k_1}, \mathbf{k_2}, \mathbf{q})$, we then find

$$\Phi_1(\mathbf{k_1}, \mathbf{q}) = \alpha_s h^2 \int d\rho \, d\tau \, \rho(1 - \rho) \left[\frac{1}{(\mathbf{q}^2 \rho^2 \tau(1 - \tau) + m^2)} \right.$$
$$\left. - \frac{1}{[(\mathbf{k_1} - \rho\mathbf{q})^2 \tau(1 - \tau) + m^2]} \right]. \quad (4.44)$$

The expression for $\Phi_2(\mathbf{k_2}, \mathbf{q})$ is obtained *mutatis mutandis*.

For zero momentum transfer the expression for the impact factors simplifies to

$$\Phi_1(\mathbf{k_1}, \mathbf{0}) = \frac{\alpha_s h^2}{6} \int d\tau \frac{\mathbf{k_1}^2 \tau(1 - \tau)}{m^2(\mathbf{k_1}^2 \tau(1 - \tau) + m^2)}, \quad (4.45)$$

with a similar expression for $\Phi_2(\mathbf{k_2}, \mathbf{0})$.

Following Balitsky & Lipatov (1978) we have organized the perturbation expansion in such a way that we consider only the leading order contribution to the impact factors and in particular we describe the meson in terms of its lowest order Fock space component (i.e. a quark and an antiquark), all higher order terms being in the quantity $f(\omega, \mathbf{k_1}, \mathbf{k_2}, \mathbf{q})$. This is a matter of choice and we could have organized the perturbation expansion differently. Indeed Mueller (1994), Chen & Mueller (1995), Nikolaev & Zakharov (1994) and Nikolaev, Zakharov & Zoller (1994a, 1994b) have considered the Fock space expansion of a heavy quark meson, starting with a quark–antiquark pair and adding any number of soft gluons. From this procedure an expression for the soft gluon contribution to the meson wavefunction can be obtained and this in turn leads to a determination of the low-x structure function of the meson which is shown to obey the BFKL equation. We shall have more to say on this way of looking at high energy scattering in Chapter 8.

A derivation of structure functions from the consideration of the sum of all possible soft gluon insertions has also been carried out by Catani, Fiorani & Marchesini (1990a,b), Catani, Fiorani, Marchesini & Oriani (1991) and Ciafaloni (1988). The application of the $t = 0$ BFKL equation in low-x deep inelastic scattering will be discussed in detail in Chapter 6.

4.5 Solution for non-zero momentum transfer

The solution of the BFKL equation for t $(= -\mathbf{q}^2)$ not equal to zero proved rather recalcitrant. Eight years elapsed from the publication of the paper by Balitsky & Lipatov (1978), in which the solution for zero momentum transfer was presented, until Lipatov (1986) solved the equation for non-zero momentum transfer. In order to do so it was first necessary to perform a two-dimensional Fourier transform to express the amplitude $f(\omega, \cdots)$ not in terms of transverse momenta $\mathbf{k_1}, \mathbf{k_2}, \mathbf{q} - \mathbf{k_1}, \mathbf{q} - \mathbf{k_2}$, but in terms of corresponding impact parameters $\mathbf{b_1}, \mathbf{b_2}, \mathbf{b_1'}, \mathbf{b_2'}$. Thus we define

$$\tilde{f}(\omega, \mathbf{b_1}, \mathbf{b_1'}, \mathbf{b_2}, \mathbf{b_2'}) = \int d^2\mathbf{k_1} d^2\mathbf{k_2} d^2\mathbf{q}$$

$$\times \left[e^{i(\mathbf{k_1} \cdot \mathbf{b_1} + (\mathbf{q} - \mathbf{k_1}) \cdot \mathbf{b_1'} - \mathbf{k_2} \cdot \mathbf{b_2} - (\mathbf{q} - \mathbf{k_2}) \cdot \mathbf{b_2'})} \frac{f(\omega, \mathbf{k_1}, \mathbf{k_2}, \mathbf{q})}{\mathbf{k_2^2}(\mathbf{k_1} - \mathbf{q})^2} \right] \quad (4.46)$$

and the BFKL equation in impact parameter space becomes

$$\omega \partial_{\mathbf{b_1}}^2 \partial_{\mathbf{b_1'}}^2 \tilde{f}(\omega, \mathbf{b_1}, \mathbf{b_1'}, \mathbf{b_2}, \mathbf{b_2'}) = (2\pi)^4 \delta^2(\mathbf{b_1} - \mathbf{b_2})\delta^2(\mathbf{b_1'} - \mathbf{b_2'})$$

$$+ \frac{\bar{\alpha}_s}{2\pi} \left\{ (2\pi)^2 \delta^2(\mathbf{b_1} - \mathbf{b_1'})(\partial_{\mathbf{b_1}} + \partial_{\mathbf{b_1'}})^2 \tilde{f}(\omega, \mathbf{b_1}, \mathbf{b_1'}, \mathbf{b_2}, \mathbf{b_2'}) \right.$$

$$+ \partial_{\mathbf{b_1}}^2 \int \frac{d^2\mathbf{c}}{(\mathbf{b_1} - \mathbf{c})^2} \left[\partial_{\mathbf{b_1'}}^2 \tilde{f}(\omega, \mathbf{c}, \mathbf{b_1'}, \mathbf{b_2}, \mathbf{b_2'}) \right.$$

$$\left. - \frac{(\mathbf{b_1} - \mathbf{b_1'})^2}{\left[(\mathbf{b_1} - \mathbf{c})^2 + (\mathbf{b_1'} - \mathbf{c})^2\right]} \partial_{\mathbf{b_1'}}^2 \tilde{f}(\omega, \mathbf{b_1}, \mathbf{b_1'}, \mathbf{b_2}, \mathbf{b_2'}) \right]$$

$$+ \partial_{\mathbf{b_1'}}^2 \int \frac{d^2\mathbf{c}}{(\mathbf{b_1'} - \mathbf{c})^2} \left[\partial_{\mathbf{b_1}}^2 \tilde{f}(\omega, \mathbf{b_1}, \mathbf{c}, \mathbf{b_2}, \mathbf{b_2'}) \right.$$

$$\left. \left. - \frac{(\mathbf{b_1} - \mathbf{b_1'})^2}{\left[(\mathbf{b_1} - \mathbf{c})^2 + (\mathbf{b_1'} - \mathbf{c})^2\right]} \partial_{\mathbf{b_1}}^2 \tilde{f}(\omega, \mathbf{b_1}, \mathbf{b_1'}, \mathbf{b_2}, \mathbf{b_2'}) \right] \right\}, \quad (4.47)$$

where $\partial_{\mathbf{b}}^2$ is the two-dimensional d'Alembertian operator with respect to the impact parameter $\mathbf{b}$. We explain how this equation is derived in the appendix to this chapter.

Once again this is a Green function equation and can be solved if we can find a complete set of eigenfunctions, $\tilde{\phi}_i(\mathbf{b}, \mathbf{b'})$, of the kernel $\tilde{\mathcal{K}}_0$, where

$$\tilde{\mathcal{K}}_0 \bullet \tilde{\phi}_i(\mathbf{b}, \mathbf{b'}) = \frac{\bar{\alpha}_s}{2\pi} \left\{ \partial_{\mathbf{b}}^2 \int \frac{d^2\mathbf{c}}{(\mathbf{b} - \mathbf{c})^2} \left[\partial_{\mathbf{b'}}^2 \tilde{\phi}_i(\mathbf{c}, \mathbf{b'}) \right. \right.$$

$$\left. - \frac{(\mathbf{b} - \mathbf{b'})^2}{\left[(\mathbf{b} - \mathbf{c})^2 + (\mathbf{b'} - \mathbf{c})^2\right]} \partial_{\mathbf{b'}}^2 \tilde{\phi}_i(\mathbf{b}, \mathbf{b'}) \right]$$

$$+ \partial_{\mathbf{b'}}^2 \int \frac{d^2\mathbf{c}}{(\mathbf{b'} - \mathbf{c})^2} \left[\partial_{\mathbf{b}}^2 \tilde{\phi}_i(\mathbf{b}, \mathbf{c}) \right.$$

$$\left. \left. - \frac{(\mathbf{b} - \mathbf{b'})^2}{\left[(\mathbf{b} - \mathbf{c})^2 + (\mathbf{b'} - \mathbf{c})^2\right]} \partial_{\mathbf{b}}^2 \tilde{\phi}_i(\mathbf{b}, \mathbf{b'}) \right] \right\} \quad (4.48)$$

(for $\mathbf{b} \neq \mathbf{b'}$).

These eigenfunctions are best described by expressing the two-dimensional vectors $\mathbf{b_i} = (b_i, \theta_i)$ in terms of complex numbers, namely,

$$b_i = b_i e^{i\theta_i},$$

so that $\partial_{\mathbf{b_i}}^2 = 4\partial^2/\partial b_i \partial b_i^*$. The eigenfunctions of $\tilde{K}_0$ are then given by

$$\tilde{\phi}_n^\nu(b, b', c) = \left(\frac{(b - b')}{(b - c)(b' - c)} \right)^n \left(\frac{|b - b'|}{|b - c||b' - c|} \right)^{1 + 2i\nu - n} \tag{4.49}$$

for any two-dimensional (transverse) vector c.

These eigenfunctions were originally identified by exploiting the two-dimensional conformal invariance of Eq.(4.48) and the fact that the expression on the right hand side of Eq.(4.49) is a representation of the two-dimensional conformal group. We shall not concern ourselves with such technicalities in this book.

If the eigenfunctions are inserted into Eq.(4.48), then, after some straightforward but tedious algebra, it can be shown that they are indeed eigenfunctions of $\tilde{K}_0$ with exactly the same eigenvalues as for the $\mathbf{q}^2 = 0$ case. In other words, the eigenvalues are also given by Eq.(4.25).

The eigenfunctions $\tilde{\phi}_n^\nu$ can be shown to obey the following completeness relation (returning to two-dimensional vector notation):

$$\sum_{n=-\infty}^{\infty} \int_{-\infty}^{\infty} d\nu \int d^2\mathbf{c}(4\nu^2 + n^2)\tilde{\phi}_n^\nu(\mathbf{b_1}, \mathbf{b_1'}, \mathbf{c})\tilde{\phi}_n^{\nu*}(\mathbf{b_2}, \mathbf{b_2'}, \mathbf{c})$$

$$= \frac{1}{4}(2\pi)^4(\mathbf{b_1} - \mathbf{b_1'})^4\delta^2(\mathbf{b_1} - \mathbf{b_2})\delta^2(\mathbf{b_1'} - \mathbf{b_2'}). \tag{4.50}$$

The derivation of this completeness relation is given in the paper by Lipatov (1986) and we refer the reader to that paper for details. We can also show quite easily that

$$\partial_{\mathbf{b_1}}^2 \partial_{\mathbf{b_1'}}^2 \tilde{\phi}_n^\nu(\mathbf{b_1}, \mathbf{b_1'}, \mathbf{c}) = \frac{(4\nu^2 + 1 - n^2)^2 + 16n^2\nu^2}{(\mathbf{b_1} - \mathbf{b_1'})^4} \tilde{\phi}_n^\nu(\mathbf{b_1}, \mathbf{b_1'}, \mathbf{c}), \tag{4.51}$$

which is useful since the operator $\partial_{\mathbf{b_1}}^2 \partial_{\mathbf{b_1'}}^2$ appears on the left hand side of Eq.(4.47).

Combining Eqs.(4.50) and (4.51) we obtain the general solution of Eq.(4.47):

$$\tilde{f}(\omega, \mathbf{b_1}, \mathbf{b_1'}, \mathbf{b_2}, \mathbf{b_2'}) = \sum_{n=-\infty}^{\infty} \int_{-\infty}^{\infty} d\nu \int d^2\mathbf{c}$$

$$\times \frac{(16\nu^2 + 4n^2)}{((4\nu^2 + 1 - n^2)^2 + 16n^2\nu^2)} \frac{\tilde{\phi}_n^\nu(\mathbf{b_1}, \mathbf{b_1'}, \mathbf{c})\tilde{\phi}_n^{\nu*}(\mathbf{b_2}, \mathbf{b_2'}, \mathbf{c})}{(\omega - \bar{\alpha}_s\chi_n(\nu))}. \tag{4.52}$$

Note that for the case where $n = \pm 1$ the integral over ν is interpreted in the sense of its principal value, namely,

$$
\int_{-\infty}^{\infty} \frac{d\nu}{\nu^2} f(\nu) = \lim_{\epsilon \to 0} \left[\int_{-\infty}^{-\epsilon} \frac{d\nu}{\nu^2} f(\nu) + \int_{\epsilon}^{\infty} \frac{d\nu}{\nu^2} f(\nu) - 2 \frac{f(0)}{\epsilon} \right].
$$

As in the case of zero momentum transfer we look for the leading singularity in ω by considering only the $n = 0$ term in the sum over n and expanding $\chi_0(\nu)$ up to quadratic order in ν. The integration over ν can then be performed. The result is rather cumbersome and we do not write it down here.

There is an important complication that arises if we wish to consider the coupling of the Pomeron to individual quarks inside the hadron. Since Eq.(4.47) is not an equation for $\tilde{f}$ but for $\partial^2_{\mathbf{b}_1} \partial^2_{\mathbf{b}'_1} \tilde{f}$, the solution we have obtained is ambiguous up to the addition of any function which is independent of *one* of $\mathbf{b}_1, \mathbf{b}'_1$ (and by symmetry any function which is independent of *one* of $\mathbf{b}_2, \mathbf{b}'_2$). In transverse momentum space, such terms give rise to ambiguities proportional to $\delta^2(\mathbf{k}_1)$ or $\delta^2(\mathbf{k}_1 - \mathbf{q})$ (and likewise $\mathbf{k}_1 \leftrightarrow \mathbf{k}_2$). These ambiguities are therefore irrelevant when we make a convolution of $f(\omega, \mathbf{k}_1, \mathbf{k}_2, \mathbf{q})$ with impact factors that vanish when any of these transverse momenta vanish.

On the other hand the Born diagram (exchange of two gluons) in impact parameter space should give a contribution to $\tilde{f}(\omega, \mathbf{b}_1, \mathbf{b}'_1, \mathbf{b}_2, \mathbf{b}'_2)$ of

$$
\tilde{f}(\omega, \mathbf{b}_1, \mathbf{b}'_1, \mathbf{b}_2, \mathbf{b}'_2)_{\text{Born}} = \frac{\pi^2}{\omega} \ln\left((\mathbf{b}_1 - \mathbf{b}_2)^2 \right) \ln\left((\mathbf{b}'_1 - \mathbf{b}'_2)^2 \right),
$$
(4.53)

whereas Eq.(4.52) in the limit $\alpha_s \to 0$ gives

$$
\tilde{f}(\omega, \mathbf{b}_1, \mathbf{b}'_1, \mathbf{b}_2, \mathbf{b}'_2)_{\alpha_s \to 0} = \frac{\pi^2}{\omega} \ln\left(\frac{(\mathbf{b}_1 - \mathbf{b}_2)^2 (\mathbf{b}'_1 - \mathbf{b}'_2)^2}{(\mathbf{b}_1 - \mathbf{b}'_2)^2 (\mathbf{b}'_1 - \mathbf{b}_2)^2} \right)
$$
$$
\times \ln\left(\frac{(\mathbf{b}_1 - \mathbf{b}_2)^2 (\mathbf{b}'_1 - \mathbf{b}'_2)^2}{(\mathbf{b}_1 - \mathbf{b}'_1)^2 (\mathbf{b}_2 - \mathbf{b}'_2)^2} \right) . (4.54)
$$

We note that Eqs.(4.53) and (4.54) differ by terms which are independent of at least one of $\mathbf{b}_1, \mathbf{b}'_1, \mathbf{b}_2, \mathbf{b}'_2$.

Mueller & Tang (1992) have pointed out that the difference between Eqs.(4.53) and (4.54) can be accounted for by replacing

$\tilde{\phi}_0^\nu(\mathbf{b}, \mathbf{b}', \mathbf{c})$ (Eq.(4.49)) by

$$\tilde{\phi}_0^\nu(\mathbf{b}, \mathbf{b}', \mathbf{c}) = \left(\frac{(\mathbf{b} - \mathbf{b}')^2}{(\mathbf{b} - \mathbf{c})^2(\mathbf{b}' - \mathbf{c})^2}\right)^{1/2+i\nu} - \left(\frac{1}{(\mathbf{b} - \mathbf{c})^2}\right)^{1/2+i\nu}$$
$$- \left(\frac{1}{(\mathbf{b}' - \mathbf{c})^2}\right)^{1/2+i\nu}. \qquad (4.55)$$

This replacement has no effect on amplitudes obtained by convolution with impact factors which vanish at zero transverse momentum, but they *do* affect the coupling of the Pomeron to individual quarks and hence are important in discussing certain diffractive dissociation processes. This has been considered in detail by Bartels *et al.* (1995) and Forshaw & Ryskin (1995).

We defer further studies of the properties of the non-forward amplitude until Chapter 7.

4.6 Deviations from 'soft' Pomeron behaviour

We have derived the hard Pomeron in (leading logarithm) perturbative QCD. It is quite distinct from the soft Pomeron of Chapter 1. Let us summarize the main differences:

1. The leading singularity of the Mellin transform is a cut and not an isolated pole. We shall return to this matter in the next chapter.

2. The position of the leading singularity gives an s-dependence $s^{\alpha_P(t)}$, where

$$\alpha_P(t) = 1 + 4\bar{\alpha}_s \ln 2.$$

This is typically much larger than the phenomenologically observed intercept of the Pomeron at $\alpha_P(0) = 1.08$ (see Donnachie & Landshoff (1992)). Moreover, $\alpha_P(t)$ is not independent of the nature of the scattering particles. This is because, in QCD, the magnitude of $\bar{\alpha}_s$ depends upon the typical size of those particles.

One of the consequences of this is that the unitarity bound of Froissart (1961) and Martin (1963) which tells us that cross-sections cannot grow with s faster than $\ln^2 s$, will be very rapidly violated. We return to the question of the restoration of unitarity in the final chapter.

3. The spectrum of singularities of the Mellin transform is the same for $\mathbf{q}^2 = 0$ as for $\mathbf{q}^2 \neq 0$, i.e. there appears to be no t-dependence of the Pomeron trajectory.

4. Factorization of the amplitude into the product of couplings of the Pomeron to the two incoming hadrons and a Pomeron amplitude only occurs *inside* an integral over transverse momentum, i.e. the integrand of the $\mathbf{k_1}, \mathbf{k_2}$ integrals factorizes into two impact factors and a Pomeron amplitude as shown in Eq.(4.36).

5. The quark-counting rule (Landshoff & Polkinghorne (1971)), which tells us that the coupling of a Pomeron to hadrons is proportional to the number of valence quarks inside the hadron, does not appear to be obeyed. This is because both diagrams of Fig. 4.4 need to be considered in order to have an impact factor which vanishes when $\mathbf{k_1} \to 0$ so that the amplitude is infra-red finite. The graph of Fig. 4.4(b) clearly violates this quark-counting rule since the two sides of the ladder couple to different quarks inside the hadron. However, if Mueller & Tang's prescription is used then it turns out that in certain kinematic regions (e.g. for large t diffractive dissociation processes) the amplitude is dominated by the contribution to the impact factor from Fig. 4.4(a) and is therefore consistent with quark counting.

Analysis of the interface between the soft and hard Pomerons within the context of QCD still presents a challenge. Nevertheless the object that we have been describing so far (the hard Pomeron) should be observable in processes for which the kinematics justifies the use of perturbation theory. We shall turn to a detailed study of the phenomenological implications of this hard Pomeron in Chapters 6 and 7.

4.7 Higher order corrections

So far, all our calculations have been performed in the leading logarithm approximation. In other words we have taken the leading term in an expansion in $1/\ln s$. In particular, we have noted that the leading term in this expansion gives a leading behaviour $s^{\omega_0}/\sqrt{\ln s}$, where ω_0 is the position of the leading singularity in the Mellin transform of the colour singlet exchange amplitude, and is $\mathcal{O}(\alpha_s)$. It is perfectly possible that the sub-leading terms could sum to give $s^{\omega_1}/(\ln s)^{3/2}$, where ω_1 is also $\mathcal{O}(\alpha_s)$. We see that, for each order in α_s, this expression is suppressed by a power of $\ln s$ relative to the leading logarithm expansion, but if $\omega_1 > \omega_0$ then the summation of the sub-leading logarithms will dominate at sufficiently large s. That this does not happen is an 'act of faith' based on the assumption that $1/\ln s$ is a good expansion parameter and that the leading term should therefore dominate at large s.

Furthermore, the problem of the violation of unitarity mentioned in the preceding section is, as pointed out by Bartels (1980), closely linked with the sub-leading logarithm contributions. The leading $\ln s$ amplitude is obtained by considering cut ladders where the only intermediate states considered are those consisting of gluons radiated off a single reggeized gluon. However, unitarity relates the imaginary part of the amplitude to the sum over *all* possible intermediate states, including those that cannot be produced via colour octet exchange. Thus the leading logarithm approximation does not lead to a unitary amplitude.

It is therefore clear that a full analysis of the sub-leading $\ln s$ contribution is very important for a complete understanding of the perturbative Pomeron.

If we look at all the places where we have made approximations valid *only* for leading logarithms: the multi-Regge kinematic regime which requires $\rho_i \gg \rho_{i+1}$, $|\lambda_{i+1}| \gg |\lambda_i|$ as we go down the ladder; the eikonal approximation for the coupling of soft gluons; the absence of fermion loops; the domination of ladders with reggeized gluons in their vertical lines; etc., we can immediately appreciate that extracting the sub-leading $\ln s$ contribution to the colour singlet exchange amplitude is a formidable task.

Nevertheless, considerable progress has been made both in the systematic calculation of the next-to-leading logarithmic corrections to colour singlet exchange and in the construction of a theory which is unitary. To detail this progress would fill another text book and so we limit ourselves here to a brief chronology of the progress that has been made. Our aim is to provide the reader with a broad overview of the area of sub-leading corrections which will provide a platform for further detailed study.

There are essentially two main lines of research which define the progress that has been made in understanding the corrections to the BFKL equation. The first line that we shall discuss is motivated by the desire to ensure that the theory be unitary, whilst the second is motivated by the need to compute all the next-to-leading logarithmic corrections to the BFKL equation.

The first attempts to correct the BFKL equation to bring it in line with unitarity date back to Bartels (1980) and Gribov, Levin & Ryskin (1983). Bartels considered the T-matrix for $m \to n$ scattering. Starting from the lowest order elements (i.e. t-channel exchange of a single reggeized gluon) one is able, using the (s-channel) unitarity relation of Eq.(1.1), to compute the matrix elements at the next order. For example, feeding the lowest order $2 \to n$ matrix element into the right hand side of Eq.(1.1) leads to the $2 \to 2$ matrix element also at lowest order (for octet exchange) or the $2 \to 2$ matrix element at the next order (for singlet exchange). The former is the bootstrap relation we used to prove the reggeization of the gluon in Chapter 3, whilst the latter is none other than the exchange of a BFKL Pomeron.[†] An iterative process can be built up, whereby the higher order corrections are computed from the lower orders in order to fulfil the demands of unitarity. The higher order corrections obtained in this way correspond to a minimal subset of higher order corrections which is determined by the requirements of unitarity. The graphs which constitute this minimal subset are those with the exchange of n reggeized gluons in the t-channel, as in Fig. 4.5, i.e. included are all those graphs which have the Reggeons interacting pairwise via the exchange of gluon rungs (the interaction being described by the BFKL kernel). Clearly, there are many other corrections which are not included in this minimal subset. For example, any graph which does not conserve the number of Reggeons in the t-channel is beyond this approximation. The transition of two Reggeons to four Reggeons has been studied in the papers by Bartels (1993a,b), Bartels & Wüsthoff (1995) and Bartels, Wüsthoff & Lipatov (1995). This work constitutes the development of the original 'fan diagram' calculations (Gribov, Levin & Ryskin (1983) and Mueller & Qiu (1986)) so as to account for

---

[†] Following Bartels, we have referred to this as the next order contribution since the even signature factor associated with the Pomeron exchange is suppressed by one power of α_s relative to the odd signature exchange of the reggeized gluon.

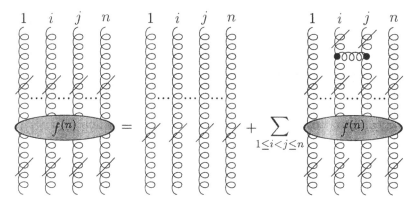

Fig. 4.5. Equation determining the evolution of the n-Reggeon state.

the full Regge kinematics.

Following Bartels (1980), Kwiecinski & Praszalowicz (1980) studied the specific case of the exchange of three reggeized gluons in an overall colour singlet state and with odd charge conjugation, i.e. the odderon. Although the integral equation describing the evolution of the n Reggeon state can easily be written (see Fig. 4.5), its solution is rather more difficult to extract. Significant progress has been made in the papers by Lipatov (1990, 1993, 1994), Kirschner (1994), Korchemsky (1995, 1996) and Faddeev & Korchemsky (1995), where a number of remarkable properties of these colour singlet Reggeon compound states have been established. We shall discuss unitarization corrections further in the final chapter.

The program of computing the next-to-leading logarithmic corrections to the BFKL equation was started in the papers by Lipatov & Fadin (1989a,b), where the leading logarithm tree level amplitudes were corrected to account for the relaxation of the Regge kinematics (i.e. strong ordering of Sudakov components) to the so-called **quasi-multi-Regge kinematics**. The radiative corrections (i.e. quark and gluon loop contributions) to the basic vertices (Reggeon–Reggeon–particle and particle–particle–Reggeon) were computed in the papers by Fadin & Fiore (1992), Fadin & Lipatov (1992, 1993), and Fadin, Fiore & Quartarolo (1994a,b). The two-loop corrections to the gluon Regge trajectory were computed by Fadin, Fiore & Quartarolo (1996) and Fadin, Fiore & Kotsky (1996). The final element which is required to complete this programme of work is to compute the amplitudes for the production of a pair of quarks or gluons in the quasi-multi-Regge kinematics (i.e. the Reggeon–Reggeon–particle–particle vertices which occur due to the relaxation of the strong ordering) and this was undertaken by Fadin & Lipatov (1996). The cancellation of infra-red divergences is expected to occur between the real and virtual graphs and has been demonstrated explicitly for the fermion contribution in Fadin & Lipatov (1996). The ultra-violet divergences, which occur due to the presence of the radiative corrections, do not cancel and are renormalized into the running of the QCD coupling.

Considerable progress in understanding the 'scale invariant' part of the sub-leading corrections has been made by White *et al.* (see e.g. Corianò & White (1995, 1996) and references therein). Their programme makes use of the simplifications which are afforded when one computes amplitudes using the t-channel analogue of Eq.(1.1) (i.e. one considers discontinuities in the t-channel rather than the s-channel). The 'Reggeon diagrams' which determine the sub-leading corrections are then straightforward to classify and, after utilizing gauge invariance (which is implemented in the form of a Ward identity) and the property that the amplitudes be infra-red finite, completely calculable. The property of infra-red finiteness leads to the fact that only the infra-red or 'scale invariant' parts of the sub-leading corrections are calculated exactly, e.g. this approach is not able to generate the radiative corrections which lead to the renormalization of the QCD coupling. However, in the lowest order the method reproduces the complete BFKL equation.

Before leaving our resumé on the progress made in computing the sub-leading corrections to the leading logarithm approximation, let us note that significant progress has been made in the construction of an effective action which can be derived directly from the original action of QCD but which is appropriate in the high energy limit. It is the hope that such an action will be useful in simplifying the calculation of multi-Reggeon Green functions due to the fact that unimportant degrees of freedom have been eliminated (recall that QCD at high energies is concerned essentially with dynamics only in the transverse plane). We refer the interested reader to the papers by Lipatov (1991, 1995), Verlinde & Verlinde (1993) and Kirschner, Lipatov & Szymanowski (1994).

4.8 Summary

• The Pomeron in leading logarithm approximation is obtained by considering colour singlet ladder diagrams whose vertical lines are reggeized gluons with couplings to the gluon rungs given by the effective vertices, Γ.

• The integral equation for the Pomeron (the BFKL equation) is obtained in the same way as the integral equation for the reggeized gluon, but with different colour factors.

• The Pomeron has even signature. In leading logarithm approximation the amplitude is purely imaginary and is suppressed by one power of α_s relative to the reggeized gluon.

• The integral equation is solved for the case of zero momentum transfer by finding the eigenfunctions of the kernel, $\mathcal{K}_0$, given by Eq.(4.18), and their corresponding eigenvalues.

• The eigenvalues are continuous, depending on a discrete variable, n, and a continuous variable, ν. This leads to a cut rather than a pole in the Mellin transform of the Pomeron amplitude, with a branch point at ω_0, given by Eq.(4.31). The leading logarithm behaviour is

$$s^{1+\omega_0}/\sqrt{\ln s}.$$

• The colour singlet exchange amplitude corresponding to the Pomeron (2 gluons $\rightarrow$ 2 gluons) is free from infra-red divergences.

• The remaining infra-red divergence (which arises from the integration over the transverse momenta of the external gluons) is removed by taking a convolution of the Pomeron with impact factors. The impact factors determine the coupling of the Pomeron to colour singlet hadrons and necessarily vanish when the transverse momentum of any of the gluon legs vanishes.

• The BFKL equation can also be solved for non-zero momentum transfer but it is first necessary to perform a two-dimensional Fourier transform and to work in impact parameter space rather than with the transverse momenta of the external gluons. The spectrum of eigenvalues is identical to that for zero momentum transfer.

• Some considerable progress has been made in calculating higher order corrections to the perturbative Pomeron, but a complete summation of all sub-leading $\ln s$ contributions has not yet been achieved.

4.9 Appendix

In this appendix we outline the derivation of Eq.(4.47) which gives the BFKL equation in impact parameter space. Firstly let us define

$$\hat{f}(\omega, \mathbf{k_1}, \mathbf{k_2}, \mathbf{q}) = \frac{f(\omega, \mathbf{k_1}, \mathbf{k_2}, \mathbf{q})}{\mathbf{k_2^2}(\mathbf{k_1} - \mathbf{q})^2},$$

so that $\tilde{f}(\omega, \mathbf{b_1}, \mathbf{b_1'}, \mathbf{b_2}, \mathbf{b_2'})$ defined in Eq.(4.46) is actually the two-dimensional Fourier transform of $\hat{f}(\omega, \mathbf{k_1}, \mathbf{k_2}, \mathbf{q})$. The BFKL equation (Eq.(4.16)) then becomes

$$\mathbf{k_1^2}(\mathbf{k_1} - \mathbf{q})^2 \omega \hat{f}(\omega, \mathbf{k_1}, \mathbf{k_2}, \mathbf{q}) \;=\; \delta^2(\mathbf{k_1} - \mathbf{k_2}) + \frac{\bar{\alpha}_s}{2\pi} \int d^2k'$$

$$\left\{ -\mathbf{q}^2 \hat{f}(\omega, \mathbf{k'}, \mathbf{k_2}, \mathbf{q}) \;+\; \frac{\mathbf{k_1^2}}{(\mathbf{k_1} - \mathbf{k'})^2} \left[(\mathbf{k'} - \mathbf{q})^2 \hat{f}(\omega, \mathbf{k'}, \mathbf{k_2}, \mathbf{q}) \right.\right.$$

$$\left. -\frac{\mathbf{k_1^2}(\mathbf{k_1} - \mathbf{q})^2}{[\mathbf{k'^2} + (\mathbf{k'} - \mathbf{k_1})^2]} \hat{f}(\omega, \mathbf{k_1}, \mathbf{k_2}, \mathbf{q}) \right]$$

$$+ \frac{(\mathbf{k_1} - \mathbf{q})^2}{(\mathbf{k_1} - \mathbf{k'})^2} \left[\mathbf{k'^2} \hat{f}(\omega, \mathbf{k'}, \mathbf{k_2}, \mathbf{q}) \right.$$

$$\left.\left. -\frac{\mathbf{k_1^2}(\mathbf{k_1} - \mathbf{q})^2}{[(\mathbf{k'} - \mathbf{q})^2 + (\mathbf{k'} - \mathbf{k_1})^2]} \hat{f}(\omega, \mathbf{k_1}, \mathbf{k_2}, \mathbf{q}) \right] \right\}. \quad (A.4.1)$$

Now we note the following Fourier transforms (F.T.):

$$-\partial_{\mathbf{b_1}}^2 \tilde{f}(\omega, \mathbf{b_1}, \mathbf{b_1'}, \mathbf{b_2}, \mathbf{b_2'}) \;=\; \text{F.T.}\left\{ \mathbf{k'^2} \hat{f}(\omega, \mathbf{k'}, \mathbf{k_2}, \mathbf{q}) \right\}$$

$$-\partial_{\mathbf{b_1'}}^2 \tilde{f}(\omega, \mathbf{b_1}, \mathbf{b_1'}, \mathbf{b_2}, \mathbf{b_2'}) \;=\; \text{F.T.}\left\{ (\mathbf{k'} - \mathbf{q})^2 \hat{f}(\omega, \mathbf{k'}, \mathbf{k_2}, \mathbf{q}) \right\}$$

(and identical expressions with $\mathbf{k'}$ replaced by $\mathbf{k_1}$), and

$$-(\partial_{\mathbf{b_1}} + \partial_{\mathbf{b_1'}})^2 \tilde{f}(\omega, \mathbf{b_1}, \mathbf{b_1'}, \mathbf{b_2}, \mathbf{b_2'}) = \text{F.T.}\left\{ \mathbf{q}^2 \hat{f}(\omega, \mathbf{k'}, \mathbf{k_2}, \mathbf{q}) \right\},$$

so that

$$-\int d^2\mathbf{k}'\mathbf{q}^2 \hat{f}(\omega,\mathbf{k}',\mathbf{k}_2,\mathbf{q})$$

is the (inverse) Fourier transform of

$$(2\pi)^2\delta^2(\mathbf{b}_1 - \mathbf{b}_1')(\partial_{\mathbf{b}_1} + \partial_{\mathbf{b}_1'})^2 \tilde{f}(\omega,\mathbf{b}_1,\mathbf{b}_1',\mathbf{b}_2,\mathbf{b}_2')$$

$((\mathbf{b}_1 - \mathbf{b}_1')$ is the impact parameter conjugate to $\mathbf{k}_1$).

We have shown in Section 4.3 that the integral

$$\int \frac{\mathbf{k}_1^2 d^2\mathbf{k}'}{(\mathbf{k}' - \mathbf{k}_1)^2[\mathbf{k}'^2 + (\mathbf{k}' - \mathbf{k}_1)^2]} = \int_0^{\mathbf{k}_1^2} \frac{d^2\mathbf{k}'}{\mathbf{k}'^2}$$

(plus integrals which cancel). This is infra-red divergent and so we regularize it by writing it as

$$\lim_{\lambda \to 0} \int_0^{\mathbf{k}_1^2} \frac{d^2\mathbf{k}'}{(\mathbf{k}'^2 + \lambda^2)} = \pi\ln\left(\frac{\mathbf{k}_1^2 + \lambda^2}{\lambda^2}\right);$$

this is the (inverse) Fourier transform of

$$\lim_{\lambda \to 0} \frac{1}{(\mathbf{b}^2 + \lambda^2)}.$$

The Fourier transform of the product of two functions $g(\mathbf{k}), h(\mathbf{k})$ is given by the convolution

$$\int d^2\mathbf{k}' g(\mathbf{k}')h(\mathbf{k}')e^{i\mathbf{b}\cdot\mathbf{k}'} = \frac{1}{(2\pi)^2}\int d^2\mathbf{c}\,\tilde{g}(\mathbf{c})\tilde{h}(\mathbf{b} - \mathbf{c}).$$

For example,

$$\int \frac{d^2\mathbf{c}}{(\mathbf{b} - \mathbf{c})^2}\tilde{g}(\mathbf{c})$$

$$= \text{ F.T.}\left\{\int d^2\mathbf{k}'\frac{\mathbf{k}_1^2}{(\mathbf{k}_1 - \mathbf{k}')^2[\mathbf{k}'^2 + ((\mathbf{k}' - \mathbf{k}_1)^2]}g(\mathbf{k}_1)\right\} \quad \text{(A.4.2)}$$

and, conversely,

$$\int d^2\mathbf{c}\frac{(\mathbf{b}_1 - \mathbf{b}_1')^2}{(\mathbf{b}_1 - \mathbf{c})^2\left[(\mathbf{b}_1 - \mathbf{c})^2 + (\mathbf{b}_1' - \mathbf{c})^2\right]}\tilde{g}(\mathbf{h}_1,\mathbf{b}_1')$$

$$= \text{ F.T.}\left\{\int \frac{d^2\mathbf{k}'}{(\mathbf{k}' - \mathbf{k}_1)^2}g(\mathbf{k}')\right\}. \quad \text{(A.4.3)}$$

(In Eq.(A.4.3) we shift the integration variable on the left hand side to $(\mathbf{c} - \mathbf{b}_1)$ and again use the fact that $\mathbf{b}_1 - \mathbf{b}_1'$ is the variable conjugate to $\mathbf{k}_1$.

Inserting $\mathbf{k}_1^2 f(\omega, \mathbf{k}_1, \mathbf{k}_2, \mathbf{q})$ or $(\mathbf{k}_1 - \mathbf{q})^2 f(\omega, \mathbf{k}_1, \mathbf{k}_2, \mathbf{q})$ for $g(\mathbf{k}_1)$ in Eq.(A.4.2) where necessary, and likewise $\mathbf{k}'^2 f(\omega, \mathbf{k}', \mathbf{k}_2, \mathbf{q})$ or $(\mathbf{k}' - \mathbf{q})^2 f(\omega, \mathbf{k}', \mathbf{k}_2, \mathbf{q})$ for $g(\mathbf{k}')$ in Eq.(A.4.3) and recalling that an extra power of $\mathbf{k}_1^2$ or $(\mathbf{k}_1 - \mathbf{q})^2$ can be obtained by acting on the left with $-\partial_{\mathbf{b}_1}^2$ or $-\partial_{\mathbf{b}_1'}^2$, respectively, the result, Eq.(4.47), follows.

Note that the product terms in Eq.(A.4.1) become convolution terms under the Fourier transform and *vice versa*.

5

From cuts to poles

As we pointed out in the preceding chapter, there are several important differences between the behaviour of the perturbative QCD Pomeron which is the solution of the BFKL equation and that of the 'soft' Pomeron predicted by Regge theory and identified in total hadronic cross-sections and differential cross-sections at small tranverse momenta. Although one might have hoped that a purely perturbative analysis of QCD would yield results which were in qualitative agreement with the behaviour of the 'soft' Pomeron, it is not surprising that the results are in fact very different. Perturbative QCD theory can only be applied reliably to Green functions in which all the momenta and their scalar products are sufficiently large. In the subsequent two chapters we shall be discussing experimental situations in which such criteria are obeyed. However, total hadronic cross-sections or differential cross-sections with low momentum transfer do not obey these criteria and we must therefore expect that non-perturbative features of QCD will play a crucial role in describing such phenomena. Unfortunately a complete analysis of the non-perturbative behaviour of QCD is outside our present grasp. Nevertheless, we can investigate the 'meeting points' of perturbative and non-perturbative QCD in order to obtain some idea of how non-perturbative effects are likely to affect the Pomeron and to what extent we may expect to be able to reproduce the behaviour of hadronic cross-sections in QCD.

One of the most striking differences between the 'soft' Pomeron approach to high energy scattering and the perturbative approach, calculated by summing the leading ln s terms to all orders, is that the Mellin transform of the scattering amplitude has a cut rather than an isolated pole. Lipatov (1986) pointed out that the origin of the cut is largely due to the fact that, in the leading logarithm derivation, the strong coupling constant, α_s, is kept fixed, whereas in QCD we know that it runs. Accounting for the running of the coupling, together with some information about the infra-red behaviour of QCD (provided by the non-perturbative sector) leads to a discrete pole singularity for the Pomeron rather than a cut. We shall begin this chapter by discussing the effect of the running of the coupling.

Before we do so a caveat is in order. The effect of the running of the coupling is a part of the corrections beyond the leading logarithm approximation which were referred to in the preceding chapter. It is, strictly speaking, inconsistent to take this into account without all the other sub-leading logarithm corrections. The hope and expectation that higher order corrections are dominated by the effect of the running of the coupling has been used before in several branches of high energy physics such as the study of infra-red renormalons or corrections to the gap equation for dynamically generated spontaneous chiral symmetry breaking in Technicolour theories. We now add the study of the BFKL Pomeron to this list.

5.1 Diffusion

At first sight it may appear unnecessary to account for the running of the coupling in the BFKL equation. The argument goes like this. The scale of typical transverse momenta involved in the (Mellin transform of the) BFKL amplitude, $f(\omega, \mathbf{k_1}, \mathbf{k_2}, \mathbf{q})$, is set by the impact factors at the top and bottom of the gluon ladder. This transverse momentum, $\mathbf{k_h}$ (we assume it is the same for both the impact factors), comes from the 'primordial' transverse momentum of partons inside the scattering hadrons. Now since the BFKL equation is infra-red safe there is no need to introduce any other momentum scale and so the integrations over transverse momenta in all sections as we go down the ladder must be dominated by $\mathbf{k} \approx \mathbf{k_h}$, and so the correct value to take for the coupling constant is simply $\alpha_s(\mathbf{k_h^2})$.

This is *almost* correct but not quite. The correct statement is that in any section of the ladder the integrand of the transverse momentum integral has a maximum at $\mathbf{k} \approx \mathbf{k_h}$, but as we go further away from the top or bottom of the ladder, where the $\mathbf{k_h}$ is set, then a wider and wider range of transverse momenta becomes significant and consequently the running of the coupling becomes important.

This broadening in the range of typical $\mathbf{k}$ values involved in the loop integrals as we move along the ladder is a diffusion effect which we will now discuss in some detail. It is an important property of the BFKL amplitude to which we shall continually return in the following chapters.

Consider the BFKL amplitude for zero momentum transfer, $F(s, \mathbf{k_1}, \mathbf{k_2}, 0)$, as a function of s (rather than its Mellin transform). The asymptotic solution is given by Eq.(4.34). To simplify our notation, let us now define

$$y = \ln \frac{s}{\mathbf{k}^2}$$

$$\tau = \ln \frac{\mathbf{k_1^2}}{\mathbf{k_2^2}}.$$

and

$$\Psi(y, \tau) = \sqrt{\mathbf{k_1^2} \mathbf{k_2^2}} F(s, \mathbf{k_1}, \mathbf{k_2}, 0).$$

For large s we may use the asymptotic solution of Eq.(4.34). In which case, $\Psi(y, \tau)$ satisfies the diffusion equation:

$$\frac{\partial \Psi(y, \tau)}{\partial y} = \omega_0 \Psi(y, \tau) + a^2 \frac{\partial^2 \Psi(y, \tau)}{\partial \tau^2}. \tag{5.1}$$

Starting from the boundary condition, $\Psi(0, \tau) = \pi \delta(\tau)$, we can solve for $\Psi(y', \tau)$. The diffusion equation tells us that as y' increases so the τ-distribution broadens and so the important range of τ-values increases.

More quantitatively, we would like to know: (a) what is the mean $\ln \mathbf{k}^2$ at some point along the ladder; (b) what is the RMS spread of the $\ln \mathbf{k}^2$ distribution at this point. To answer these questions we need first to appreciate that[†]

$$F(s, \mathbf{k_1}, \mathbf{k_2}, 0) = \int d^2 \mathbf{k}' F(s', \mathbf{k_1}, \mathbf{k}', 0) F(s/s', \mathbf{k}', \mathbf{k_2}, 0), \tag{5.2}$$

for arbitrary $s' \leq s$, i.e. we can view the BFKL amplitude as a convolution of two other BFKL amplitudes with an *arbitrary* partitioning of the total energy s. We define $y' = \ln s'/\mathbf{k}^2$. For a

---

[†] This can be seen by inverting the Mellin transform of Eq.(4.28) and using the orthonormality relations of the eigenfunctions.

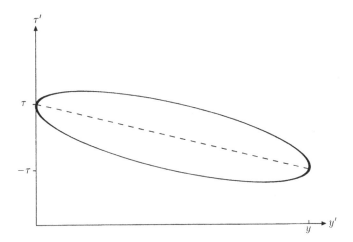

Fig. 5.1. Diffusion in the τ'–y plane.

given y' we can now ask for the mean of

$$\tau' \equiv \ln \frac{\mathbf{k}'^2}{\sqrt{\mathbf{k}_1^2 \mathbf{k}_2^2}}.$$

This is what we mean when we ask for the typical transverse momentum at some point along the ladder. It is a simple matter of Gaussian integration to compute

$$\begin{aligned}
\langle \tau' \rangle &= \int d^2 \mathbf{k}' \tau' \frac{F(s', \mathbf{k}_1, \mathbf{k}', 0) F(s/s', \mathbf{k}', \mathbf{k}_2, 0)}{F(s, \mathbf{k}_1, \mathbf{k}_2, 0)} \\
&= \frac{\tau}{2} \left(1 - 2\frac{y'}{y} \right).
\end{aligned} \tag{5.3}$$

The RMS deviation, σ, is similarly computed:

$$\begin{aligned}
\sigma^2 &= \int d^2 \mathbf{k}' (\tau' - \langle \tau' \rangle)^2 \frac{F(s', \mathbf{k}_1, \mathbf{k}', 0) F(s/s', \mathbf{k}', \mathbf{k}_2, 0)}{F(s, \mathbf{k}_1, \mathbf{k}_2, 0)} \\
&= 2a^2 y' \left(1 - \frac{y'}{y} \right).
\end{aligned} \tag{5.4}$$

In Fig. 5.1 we show a plot which illustrates the diffusion in τ'. The dotted straight line represents $\langle \tau' \rangle$ whilst the solid curves are of the functions $\langle \tau' \rangle \pm \sigma$, i.e. they represent the RMS deviations about the mean.

The axis of the 'cigar' is tilted since we chose $\tau \neq 0$, i.e. the virtualities of the external gluons are not equal. In order that we can trust a perturbative calculation, it had better be that the cigar does not dip (or tip!) too far into the region of $\mathbf{k}'^2 \sim \Lambda_{QCD}^2$. Remember we need to convolute $F(s, \mathbf{k_1}, \mathbf{k_2}, 0)$ with the relevant impact factors to obtain physical cross-sections. The avoidance of diffusion into the infra-red region is equivalent to demanding that the impact factors $\Phi_i(\mathbf{k_i})/\mathbf{k_i^2}$ be peaked at large $\mathbf{k_i^2}$.

We have just seen that even if we pick the impact factors so that the axis of the cigar is horizontal (i.e. $\tau = 0$), we still have to worry about diffusion. It is therefore more sensible to conclude that the value of α_s which should be used in the BFKL equation is $\alpha_s(\mathbf{k}'^2)$ rather than a fixed value. However, we must remember that the BFKL equation involves an integral over transverse momenta from zero upwards and hence, for sufficiently small arguments, the running coupling becomes far too large for perturbation theory to be valid. It is therefore necessary to freeze the coupling below a certain magnitude of transverse momentum (or perform some other regulating procedure). Of course this is a phenomenological procedure without any fundamental basis in QCD. Nevertheless, we now consider how to deal with such a running coupling, at least within a reasonable approximation.

5.2 Accounting for the running of the coupling

In order to solve the BFKL equation for running coupling[†] we need to find the solutions of the eigenvalue equation:

$$\bar{\alpha}_s(\mathbf{k}^2) \int \frac{d^2\mathbf{k}'}{(\mathbf{k} - \mathbf{k}')^2} \left[\phi_i(\mathbf{k}') - \frac{\mathbf{k}^2}{[\mathbf{k}'^2 + (\mathbf{k} - \mathbf{k}')^2]} \phi_i(\mathbf{k}) \right] = \lambda_i \phi_i(\mathbf{k}). \tag{5.5}$$

The running coupling, $\bar{\alpha}_s(\mathbf{k}^2)$, is given (to leading order) by

$$\bar{\alpha}_s(\mathbf{k}^2) = \frac{4N}{\beta_0 \xi}, \tag{5.6}$$

where $\xi = \ln \mathbf{k}^2/\Lambda_{QCD}^2$ and $\beta_0 = 11N/3 - 2n_f/3$ for n_f light flavours (we shall sometimes write this as $\bar{\alpha}_s(\xi)$). We could have

---

[†] For the time being we continue to work at zero momentum transfer, deferring studies of large momentum transfer to the next section.

taken the running coupling *inside* the integral and written it as $\bar{\alpha}_s(\mathbf{k'}^2)$ so that it runs with the integrated transverse momentum. The difference between the two choices depends on whether we take the coupling at a particular rung of the ladder to be controlled by the momentum of the gluon above or below that rung. Strictly speaking, one should take the maximum of the two, i.e. $\bar{\alpha}_s(\max(\mathbf{k}^2, \mathbf{k'}^2))$, but this is not necessary since the transverse momenta of two adjacent sections of the ladder are indeed of the same order.

Equation (5.5) cannot be solved analytically in the same way as we did for the case of a fixed coupling constant. There are two possible approaches to finding an approximate solution. The first is to approximate the integral equation by a large matrix (by discretizing the transverse momentum) and finding the eigenvalues and eigenvectors numerically. This was done by Daniell & Ross (1989). The other is to try analytic approximations, which is the method used by Lipatov (1986) that we discuss here.

The method used is similar to the WKB approximation for solving Schrödinger's equation, in which good approximations are found in different regions and these are matched at the turning points. Once more we restrict ourselves to the azimuthally symmetric solution $n = 0$. Motivated by the fact that for fixed coupling constant the eigenfunctions are

$$\phi_\nu^0 \sim \frac{1}{\sqrt{\mathbf{k}^2}} \exp\left(i\nu\xi\right), \tag{5.7}$$

with eigenvalues $\bar{\alpha}_s \chi_0(\nu)$, we try a solution

$$\frac{C(\xi)}{\sqrt{\mathbf{k}^2}} \exp\left(\pm i \int^\xi d\xi' \nu(\xi')\right) \tag{5.8}$$

for the eigenfunction with eigenvalue λ_i. Now, ν is treated as a function and it is related to the inverse of the function χ_0 such that,

$$\chi_0(\nu(\xi)) = \lambda_i/\bar{\alpha}_s(\xi). \tag{5.9}$$

Equation (5.8) reduces to Eq.(5.7) with $C(\xi)$ set to a constant if we take $\bar{\alpha}_s$ to be fixed. Equation (5.8) will be a good approximation as long as the function ν does not vary too much with ξ. We can obtain a good approximation for the prefactor $C(\xi)$ by inserting Eq.(5.8) into the eigenvalue equation (Eq.(5.5)) expressed as a differential equation, i.e.

$$\bar{\alpha}_s(\mathbf{k}^2)\chi_0\left(-i\frac{\partial}{\partial\xi}\right)C(\xi)\exp\left(\pm i\int^\xi d\xi'\nu(\xi')\right)$$

$$= \lambda_i\, C(\xi)\exp\left(\pm i\int^\xi d\xi'\nu(\xi')\right). \quad (5.10)$$

Assuming that $C(\xi)$ and $\nu(\xi)$ are slowly varying functions so that we may neglect second and higher order derivatives, Eq.(5.10) is satisfied provided

$$\chi_0'(\nu(\xi))\,C'(\xi) + \frac{1}{2}\chi_0''(\nu(\xi))\nu'(\xi)C(\xi) = 0, \quad (5.11)$$

which is solved by

$$C(\xi) \propto \frac{1}{\sqrt{|\chi_0'(\nu(\xi))|}}. \quad (5.12)$$

More precisely if $\nu^n(\xi)$ is the nth derivative of the function ν, then the approximation is good so long as

$$\nu^n(\xi) \ll (\nu(\xi))^n \quad (5.13)$$

(for $n \geq 1$). This condition may be true for some regions of the integration variable ξ', but it cannot be valid throughout. This is because the function $\nu(\xi)$ has a zero when $\chi_0 = 4\ln 2$, which occurs at some critical value of ξ, depending on the eigenvalue λ_i,

$$\xi_c = \frac{16N\ln 2}{\beta_0\lambda_i} \quad (5.14)$$

and its derivative becomes infinite at that point. Near $\xi = \xi_c$, ν may be approximated using Eqs. (4.29), (5.6), (5.9) and (5.14) as

$$\nu \approx \left(\frac{\lambda_i\beta_0}{56N\zeta(3)}\right)^{1/2}\sqrt{(\xi_c - \xi)}. \quad (5.15)$$

For values of ξ larger than ξ_c, ν is imaginary, and from Eq.(5.8) we see that the eigenfunction is no longer an oscillating function of ξ, but an exponentially decreasing function:

$$\phi(\mathbf{k}) = \eta\frac{1}{\sqrt{|\chi_0'(\nu(\xi))|\mathbf{k}^2}}\exp\left(-\int_{\xi_c}^\xi d\xi'|\nu(\xi')|\right), \quad (5.16)$$

where η is a (as yet undetermined) phase. Once again Eq.(5.16) is only valid away from the branch point where the inequality (5.13) is expected to hold. There is also an exponentially increasing solution (the function $\nu(\xi)$ has two branches for $\xi > \xi_c$), but like all good physicists we throw this away since an exponentially increasing eigenfunction is not physically acceptable.

What about the 'forbidden' region $\xi \approx \xi_c$. Since ν is small in this region, we may use the expansion Eq.(4.29) to expand χ_0 up to quadratic order and rewrite the integral equation as a second order differential equation:

$$\frac{4N}{\beta_0 \xi}\left(4\ln 2 + 14\zeta(3)\frac{\partial^2}{\partial \xi^2}\right)\phi_i(\mathbf{k}) = \lambda_i \phi_i(\mathbf{k}). \tag{5.17}$$

Again using Eq.(5.14), rearranging terms and changing variables from ξ to z where

$$z = \left(\frac{\beta_0 \lambda_i}{56 N \zeta(3)}\right)^{1/3}(\xi - \xi_c), \tag{5.18}$$

this equation becomes

$$\frac{\partial^2 \phi_i}{\partial z^2} - z\phi_i = 0, \tag{5.19}$$

which is Airy's equation. There exists a solution, Ai(z), which has the following asymptotic forms:

$$\text{Ai}(z) \;\rightarrow\; \frac{1}{\sqrt{\pi}|z|^{1/4}}\sin\left(\frac{2}{3}|z|^{3/2} + \frac{\pi}{4}\right), \quad z \longrightarrow -\infty$$

$$\text{Ai}(z) \;\rightarrow\; \frac{1}{2\sqrt{\pi}z^{1/4}}\exp\left(-\frac{2}{3}z^{3/2}\right), \quad z \rightarrow \infty$$

(there is also a solution which grows exponentially as $|z| \rightarrow \infty$ which corresponds to the unphysical discarded solution).

Now for sufficiently small ν where the approximation Eq.(5.15) is valid, we have

$$\frac{2}{3}|z|^{3/2} \approx \left|\int_{\xi_c}^{\xi} d\xi' \nu(\xi')\right|$$

and from Eqs.(4.29), (5.15) and (5.18),

$$|z|^{1/4}\sqrt{28\zeta(3)} = \sqrt{|\chi_0'|}\left[\frac{56N\zeta(3)}{\beta_0 \lambda_i}\right]^{1/6}.$$

Therefore we can match the region where the Airy function is a good approximation to the regions $\xi \ll \xi_c$ (oscillatory solution) and $\xi \gg \xi_c$ (exponentially decaying solution) and, furthermore, this matching uniquely determines the phase of the oscillatory solution. We may therefore write the approximate solution (up to an overall constant) for all values of ξ:

$$\phi_i(\mathbf{k}) \approx \sqrt{\frac{28\zeta(3)}{|\chi_0'(\nu(\xi))||\mathbf{k}^2}} \sin\left(\int_\xi^{\xi_c} d\xi' \nu(\xi') + \frac{\pi}{4}\right), \quad \xi \ll \xi_c,$$

$$\phi_i(\mathbf{k}) \approx \sqrt{\frac{\pi}{\mathbf{k}^2}} \left(\frac{\lambda_i \beta_0}{56N\zeta(3)}\right)^{-1/6} \mathrm{Ai}\left(\left(\frac{\beta_0 \lambda_i}{56N\zeta(3)}\right)^{1/3} (\xi - \xi_c)\right),$$

$$\xi \sim \xi_c,$$

$$\phi_i(\mathbf{k}) \approx \sqrt{\frac{7\zeta(3)}{|\chi_0'(\nu(\xi))||\mathbf{k}^2}} \exp\left(-\int_{\xi_c}^\xi d\xi' |\nu(\xi')|\right), \quad \xi \gg \xi_c. \quad (5.20)$$

We have introduced one mass scale Λ_{QCD} and this has allowed us to fix the phase of the oscillating solution at the turning point ξ_c. Now we need one more ingredient in order to compute the eigenvalues (or equivalently $\nu(\xi)$). We need to assume that the phase is fixed to some angle ϑ at some value of the (logarithm of the) transverse momentum, ξ_0. For sufficiently large momentum transfer (rather than the zero momentum transfer case that we are considering here) this second scale is provided by the momentum transfer, t, as we shall discuss below. For small or zero momentum transfer processes the value of the phase at the infra-red scale (ξ_0) must be provided by the infra-red features of QCD and cannot be attained from perturbation theory. In other words, we are going beyond perturbation theory in assuming the existence of this infrared scale which characterizes the non-perturbative behaviour of the gluons in the ladder. Now we have two scales at which the phase is fixed and in analogy with the WKB approximation for solving the Schrödinger equation this sets conditions which can only be met by certain eigenvalues. In this case matching the phase at ξ_0 gives

$$\vartheta = \int_{\xi_0}^{\xi_c} d\xi' \nu(\xi') + \frac{\pi}{4} + (i-1)\pi, \quad (5.21)$$

with i a positive integer. A typical solution is shown in Fig. 5.2. The phase is fixed at the point $\xi = \xi_0$ to the value ϑ and is also fixed in region II, so that the Airy function solution in region III matches the oscillatory solution (region II) (up to a multiple of π) and the exponentially decaying solution (region IV). Below ξ_0 (region I) the solution is dominated by the (non-perturbative)

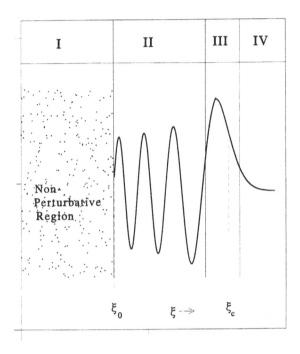

Fig. 5.2. An eigenfunction $\phi_i(\mathbf{k})$ in different regions of ξ (see Eq.(5.20)). Region I is the infra-red region dominated by non-perturbative behaviour. Region II is the oscillatory region ($\xi \ll \xi_c$). Region III is the region given by the Airy function ($\xi \sim \xi_c$) and region IV is the decay region ($\xi \gg \xi_c$). (We have chosen one of the lower eigenfunctions so that the oscillating region can be seen clearly.)

infra-red features of QCD. Now recall that ξ_c depends on the eigen-value λ_i. Thus *only certain discrete values of λ_i can be solutions of Eq.(5.21) and hence we find a discrete spectrum of the integral operator $\mathcal{K}_0$.* This means that the Mellin transform is given by a sum of isolated poles, the leading one being the solution of Eq.(5.21) with $i = 1$ and is identified as the Pomeron pole. For ξ_0 large enough for us to assume that ν is always small between ξ_0

and ξ_c we can use Eq.(5.15) for ν. It then follows that

$$\int_{\xi_0}^{\xi_c} \nu(\xi')d\xi' = \left(\frac{\lambda_i\beta_0}{56N\zeta(3)}\right)^{1/2} \frac{2}{3}(\xi_c - \xi_0)^{3/2}.$$

We can now calculate the first correction to the location of the Pomeron pole in terms of the phase angle, ϑ, i.e.

$$\alpha_P(0) \approx 1 + 4\ln 2 \frac{\bar{\alpha}_s(\xi_0)}{N} \tag{5.22}$$

$$\times \left\{1 - \left(\frac{\beta_0\bar{\alpha}_s(\xi_0)}{4N}\right)^{2/3} \left[\frac{7\zeta(3)}{2\ln 2}\right]^{1/3} \left(\frac{3(\vartheta - \pi/4)}{2}\right)^{2/3}\right\}.$$

Unfortunately, this is only the first term in a slowly convergent series and provides a very poor approximation over sensible values of ξ_0. For a full numerical solution of Eq.(5.21) we refer to the literature (Hancock & Ross (1992)). We shall shortly turn to a study of the running coupling in the case of large momentum transfer. In this case the phase is fixed by perturbation theory and we are able to quantify the location of the Pomeron (and sub-leading) poles.

However, before leaving the $t = 0$ case we wish to remark that (despite the fact that an exact analytic solution of the BFKL equation with running coupling is not possible) Collins & Kwiecinski (1989) have established upper and lower limits for the intercept of the leading trajectory. They found that if the running of the coupling is 'frozen' at some infra-red scale, $k_0^2 = \Lambda_{QCD}^2 \exp \xi_0$, then the intercept obeys the inequalities

$$1 + 1.2\bar{\alpha}_s(k_0^2) \leq \alpha_P(0) \leq 1 + 4\ln 2\,\bar{\alpha}_s(k_0^2). \tag{5.23}$$

5.3 Large momentum transfer

For non-zero momentum transfer, we proceed in the same way except that we work in impact parameter space and the eigenfunctions are functions of $\mathbf{b}, \mathbf{b}', \mathbf{c}$. From Eq.(4.49) we see that the quantity which is raised to the power $i\nu$ is

$$\left(\frac{(\mathbf{b} - \mathbf{b}')^2}{((\mathbf{b} - \mathbf{c})^2(\mathbf{b}' - \mathbf{c})^2)}\right),$$

so we replace $\mathbf{k}^2$ in the preceding section with this expression for the argument of the running of the coupling. The impact parameter $\mathbf{c}$ is integrated over and to the approximation to which we are working (leading order in the running) we can set $\mathbf{c} = \mathbf{0}$ inside the running coupling and take the quantity ξ to be

$$\xi = \ln \left(\frac{(\mathbf{b} - \mathbf{b}')^2}{\mathbf{b}^2 \mathbf{b}'^2 \Lambda_{QCD}^2} \right).$$

If $\mathbf{b} \ll \mathbf{b}'$ or $\mathbf{b}' \ll \mathbf{b}$ then the coupling is controlled by the smaller of the two impact parameters, as one would expect.

As discussed by Kirschner & Lipatov (1990) it turns out to be convenient to work in the 'mixed representation' where we keep explicit the dependence upon the momentum transfer $\mathbf{q}^2$. In other words we invert one of the two Fourier transforms that were performed to get from an eigenfunction which depended on $\mathbf{k}$ and $\mathbf{q} - \mathbf{k}$ to $\mathbf{b}$ and $\mathbf{b}'$. More precisely, we perform the inverse Fourier transform of the right hand side of Eq.(4.49) in the variable $\mathbf{b} + \mathbf{b}'$ which is conjugate to $\mathbf{q}$ and keep the remaining combination $\hat{\mathbf{b}} \equiv \mathbf{b} - \mathbf{b}'$. This leads to an eigenfunction which is a function of $\mathbf{q}$ and $\hat{\mathbf{b}}$. In this case the running of the coupling is controlled by the larger of $\mathbf{q}^2$ and $1/\hat{\mathbf{b}}^2$. When $\hat{\mathbf{b}}$ becomes larger than $1/\mathbf{q}^2$ the coupling stops running and is 'frozen' at $\alpha_s(\mathbf{q}^2)$.[†] For larger values of $\hat{\mathbf{b}}$ the solution continues to oscillate but with a fixed (angular) frequency ν_0, where

$$\lambda_i = \bar{\alpha}_s(\mathbf{q}^2) \chi_n(\nu_0). \tag{5.24}$$

The Fourier transform is straightforward but tedious. The details are given by Lipatov & Kirschner (1990). The result is

$$\phi_i(\mathbf{q}, \hat{\mathbf{b}}) \propto \sin \left(\frac{\pi}{2} - \nu \ln \left(\hat{\mathbf{b}}^2 \mathbf{q}^2 / 4 \right) + \frac{\delta(n, \nu)}{2} + n\theta_b \right) + \mathcal{O}(\hat{\mathbf{b}}\mathbf{q}), \tag{5.25}$$

where θ_b is the angle between $\hat{\mathbf{b}}$ and some fixed direction, and the phase $\delta(n, \nu)$ is given by

$$e^{i\delta(n,\nu)} = \frac{\Gamma^2((n+1)/2 + i\nu)\Gamma(n+1-2i\nu)\Gamma(-2i\nu)}{\Gamma^2((n+1)/2 - i\nu)\Gamma(n+1+2i\nu)\Gamma(2i\nu)}. \tag{5.26}$$

Equation (5.25) is valid in the region $\hat{\mathbf{b}}\mathbf{q} \ll 1$, and we have not written down the constant since it is only the phase matching that is important for the determination of the permitted eigenvalues.

[†] We assume that $\mathbf{q}^2$ is sufficiently large that $\alpha_s(\mathbf{q}^2)$ is small enough to be a valid expansion parameter in perturbation theory.

In this case we set ξ to

$$\xi = -\ln\left(4\hat{\mathbf{b}}^2\Lambda_{QCD}^2\right)$$

and once again for $\hat{\mathbf{b}} < 1/\mathbf{q}$ the coupling runs as

$$\bar{\alpha}_s(\hat{\mathbf{b}}) = \frac{4N}{\beta_0\xi}$$

and we have to replace $\nu\ln(\hat{\mathbf{b}}^2)$ in Eq.(5.25) by

$$\int^\xi \nu(\xi')d\xi'.$$

For sufficiently small values of $\hat{\mathbf{b}}$ (where $\xi > \xi_c$), ν becomes imaginary and we obtain a solution which decays exponentially with $\hat{\mathbf{b}}$. The region $\nu \approx 0$ can again be solved in terms of an Airy function and the matching of the phase tells us that for $\xi \ll \xi_c$ we have

$$\phi_i(\mathbf{q}, \hat{\mathbf{b}}) \propto \sin\left(\frac{\pi}{4} + \int_\xi^{\xi_c} \nu(\xi')d\xi'\right). \qquad (5.27)$$

Now for consistency we must match the phases in Eqs.(5.25) and (5.27) which puts a constraint on the allowed values for ξ_c and consequently also on the allowed eigenvalues, λ_i. We will impose this phase matching at $\xi = \xi_0$, where

$$\xi_0 = \ln\left(\frac{\mathbf{q}^2}{\Lambda_{QCD}^2}\right) = \frac{4N}{\beta_0\bar{\alpha}_s(\mathbf{q}^2)}. \qquad (5.28)$$

At this point $\hat{\mathbf{b}}^2 = \hat{\mathbf{b}}_0^2 = 4/\mathbf{q}^2$ and the coupling freezes (i.e. for larger impact parameters than $\hat{\mathbf{b}}_0$ the coupling is determined by $\mathbf{q}^2$ and not $\hat{\mathbf{b}}^2$). Strictly, we cannot push Eq.(5.25) as far as this because there are corrections of order $\hat{\mathbf{b}}\mathbf{q}$. However, since the coupling varies only logarithmically with $\hat{\mathbf{b}}$ we can go to a value of $\hat{\mathbf{b}}$ where $\hat{\mathbf{b}}\mathbf{q}$ is still small but $\alpha_s(\hat{\mathbf{b}}) \approx \alpha_s(\mathbf{q})$. Again we confine ourselves to the azimuthally symmetric solution ($n = 0$). Setting $\hat{\mathbf{b}}$ to $\hat{\mathbf{b}}_0$ in Eq.(5.25) (where $\ln(\hat{\mathbf{b}}^2\mathbf{q}^2/4) = 0$) and in Eq.(5.27) and matching the phases we obtain

$$\int_{\xi_0}^{\xi_c} \nu(\xi')d\xi' = \frac{(1+4i)\pi}{4} + \frac{\delta(0,\nu_0)}{2}. \qquad (5.29)$$

This fixes the allowed eigenvalues (it is directly analogous to Eq.(5.21) for the $t = 0$ case) in terms of a *perturbative* phase,

$\delta(0, \nu_0)$ ($\approx \pi$ for small ν_0). As before, for large enough $\mathbf{q}^2$, we can find the approximate solution:

$$\lambda_i \approx 4 \ln 2 \, \bar{\alpha}_s(\mathbf{q}^2)$$

$$\times \left\{ 1 - \left(\frac{\beta_0 \alpha_s(\mathbf{q}^2)}{4\pi} \right)^{2/3} \left[\frac{7\zeta(3)}{2 \ln 2} \right]^{1/3} \left(\frac{3\pi(i + 3/4)}{2} \right)^{2/3} \right\}. \quad (5.30)$$

So again, running the coupling has discretized the cut to a semi-infinite series of poles (again the analytic result is a poor approximation for attainable values of t). As we shall see in Figs. 5.4 and 5.5, the leading pole is shifted significantly downwards after taking asymptotic freedom into account. However, it is still too large to account for the behaviour of hadronic total cross-sections. Moreover, the trajectory is very flat in t (we will discuss the t-dependence further in Chapter 7), which is not consistent with (for example) the observed shrinkage of the forward diffraction peak. We can conclude, therefore, that although it is indeed possible to obtain an isolated Pomeron trajectory purely from perturbative considerations (for sufficiently large values of $\mathbf{q}^2$), non-perturbative effects are likely to be essential if we are to have any chance of reproducing the Pomeron identified in the study of soft hadron physics.

5.4 The Landshoff–Nachtmann model

We shall spend the rest of this chapter discussing various attempts that have been made to incorporate non-perturbative effects into the construction of the Pomeron in the hope of reproducing at least some of the phenomenological properties of the 'soft' Pomeron.

There are two orthogonal approaches to this. In the first approach it is assumed that the 'hard' Pomeron that we have been considering so far is heavily attenuated at small transverse momenta so that it becomes subdominant and the 'soft' Pomeron, an entirely different object which has nothing to do with gluon ladders and belongs completely to the non-perturbative realm of QCD, takes over as the dominant contribution to diffractive processes. In the second approach the 'hard' Pomeron converts smoothly into the 'soft' Pomeron at sufficiently low transverse momenta. Whereas the first approach provides an adequate explanation of why we have so far failed to reproduce any of the phenomenological properties of the 'soft' Pomeron, it offers no explanation of how this 'soft' Pomeron might arise from QCD. The second approach is more optimistic, although, as we shall see, it has so far only made very small steps towards its ultimate goal of providing a complete description of the 'soft' Pomeron.

Landshoff & Nachtmann (1987) developed a model based firmly on the Low–Nussinov picture, namely the exchange of two gluons in a colour singlet state. However, they argued that since the 'soft' Pomeron is very much controlled by the non-perturbative (infrared) aspects of QCD, one should not expect the exchanged gluons to have a propagator which at low k^2 behaves like

$$\frac{1}{k^2},$$

particularly as gluons are supposed to be confined and so the propagator cannot have a pole. These non-perturbative gluons would have a propagator, $D_{np}(k^2)$, with a much softer k^2-dependence. This non-perturbative propagator can be related to the vacuum expectation value of the square of the gluon operator:

$$\langle 0| : G_{\mu\nu}(x)G^{\mu\nu}(x) : |0\rangle = -i \int \frac{d^4k}{(2\pi)^4} 6k^2 D_{np}(k^2). \qquad (5.31)$$

In order for the integral on the right hand side of Eq.(5.31) to converge it is necessary that $D_{np}(k^2)$ falls with increasing k^2 at least as fast as $1/k^6$ (the perturbative propagator takes over at large k^2). Therefore, from dimensional analysis the non-perturbative propagator must depend on some length scale, a, provided by the infra-red region of QCD.

Landshoff and Nachtmann were concerned with the problem of quark counting in the coupling of the Pomeron to hadrons (e.g. the Pomeron coupling to a baryon with three valence quarks is depicted in Fig. 5.3). A straightforward calculation shows that the contribution from the graph of Fig. 5.3(a), where both gluons couple to the same quark, dominates over the contribution from the graph in Fig. 5.3(b) provided a is small compared with the typical hadron radius, R. If this condition can be achieved then the quark-counting rule follows with corrections of order a^2/R^2.

The model was confined to the consideration of an Abelian

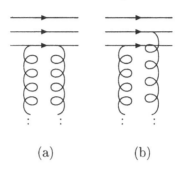

(a)   (b)

Fig. 5.3. Graphs contributing to the coupling of the Pomeron to
the three valence quarks of a baryon. Graphs of type (a) must dominate those of type (b) in order to reproduce the quark-counting
rule.

gauge theory to describe the gluons and did not address the question of obtaining the Pomeron trajectory through gluon ladders.
In a non-Abelian theory one would expect the scales a and R to be
of the same order of magnitude since they are both generated by
the same mechanism, i.e. the infra-red properties of QCD. Nevertheless, factors of 2 and π would certainly arise and it is perfectly
possible that the non-perturbative gluon propagator does behave
(at least qualitatively) in the manner suggested by Landshoff and
Nachtmann and that a is sufficiently small compared with R to
account for the observed quark-counting rule.

5.5 The effect of non-perturbative propagators

The Landshoff–Nachtmann model described above immediately
poses two important questions for those who wish to relate the
'soft' Pomeron to QCD.

1. Can non-perturbative propagators with the required low momentum properties be extracted from QCD?
2. Do gluon ladders with non-perturbative propagators for the
vertical gluons simulate the 'soft' Pomeron?

There have been several attempts to extract soft gluon propagators from various non-perturbative approaches to QCD, varying from lattice techniques to solutions of the Dyson–Schwinger

equations. Some of these investigations do indeed give propagators which are finite as $k^2 \to 0$ or at least have a softer singularity than a pole – others give propagators which have an even steeper singularity as $k^2 \to 0$ and such behaviour has been hailed as a signal for a confining gluon potential. Recently Büttner & Pennington (1995) have argued that, at least in the Landau gauge, a propagator with a small momentum behaviour softer than $1/k^2$ is inconsistent with the Dyson–Schwinger equation. They argue that the Pomeron cannot be explained in terms of the Landshoff–Nachtmann model and that its behaviour is controlled by the coupling of soft gluons to off-shell quarks inside the hadron.

Notwithstanding this, we shall investigate the effect of soft propagators, $D(k^2)$, which do *not* have a pole at $k^2 = 0$ (i.e. propagators which represent confined as opposed to confining gluons) but which, for large k^2, reduce to the usual perturbative propagators.

A complete non-perturbative treatment of the Pomeron would require knowledge about all the gluon Green functions, not just the propagator. Clearly, this is impossible and so we have to compromise. One possible approach is to make the assumption that the non-perturbative features of QCD manifest themselves mainly by the effect of the propagators for soft gluons, whereas for the vertices we may continue to use the perturbative expressions. An approach along these lines is that used by Hancock & Ross (1992, 1993) in which such non-perturbative propagators are inserted directly into the BFKL equation (with running coupling). Thus, for example, at zero momentum transfer the kernel, $\mathcal{K}_0$, of Eq.(4.18) is replaced by

$$
\mathcal{K}_0 \bullet f(\omega, \mathbf{k_1}, \mathbf{k_2}, \mathbf{q}) = \frac{\bar{\alpha}_s(\mathbf{k_1^2})}{\pi} \int d^2\mathbf{k}' D((\mathbf{k_1} - \mathbf{k}')^2)
$$

$$
\times \left[f(\omega, \mathbf{k}', \mathbf{k_2}, \mathbf{q}) \right.
$$

$$
\left. - \frac{D(\mathbf{k_1^2})D(\mathbf{k'^2})D((\mathbf{k_1} - \mathbf{k}')^2)}{D(\mathbf{k'^2}) + D((\mathbf{k_1} - \mathbf{k}')^2)} f(\omega, \mathbf{k_1}, \mathbf{k_2}, \mathbf{q}) \right]. \quad (5.32)
$$

Clearly, the eigenvalues of this operator have to be found by numerical techniques which involve discretizing the transverse momenta $\mathbf{k_1}$ and $\mathbf{k}'$ and diagonalizing the resulting matrix. The eigenvalues are discrete since there is an 'ultra-violet' scale, Λ_{QCD},

encoded in the running of the coupling, as well as an infra-red
scale, a, contained in the non-perturbative gluon propagator.[†] In
other words the infra-red scale is set to $\xi_0 = -\ln(a^2\Lambda^2_{QCD})$ and
the fact that a discrete spectrum of eigenvalues is obtained means
that the infra-red behaviour which has been introduced by replacing the propagators with non-perturbative propagators fixes the
phase ϑ (see Eq.(5.21)) at this infra-red scale, although it is difficult to understand from analytic considerations exactly how this
phase fixing mechanism works.

It turns out that the leading eigenvalue for zero momentum
transfer (i.e. the intercept of the Pomeron) does not depend on
the exact nature of the non-perturbative propagator but only on
the infra-red scale. Therefore, in Fig. 5.4 we take the simplest
possible example in which it is assumed that the infra-red effects introduce an effective mass $1/a$ for the gluon (i.e. we take
$D(\mathbf{k}^2) = a^2/(1 + a^2\mathbf{k}^2)$), and plot the intercept of the Pomeron
against $1/a$. We observe that there is a reduction of this intercept
as the effective mass is increased. The intercept is still a long way
from the observed value of 1.08 for the 'soft' Pomeron, but it is
clear that this, albeit naive, attempt to take non-perturbative effects into consideration has the effect of pushing the intercept in
the right direction.

One can also insert non-perturbative propagators into the
BFKL equation for non-zero momentum transfer and solve numerically. The result of such a procedure (taking $1/a$ to be 0.25 GeV)
for the leading trajectory and first two sub-leading trajectories is
shown in Fig. 5.5. We also show (dashed lines) the result of the
purely perturbative trajectories discussed in the preceding section (i.e. the solutions of Eq.(5.30)). We note that these perturbative solutions have a very small slope indicating a very small
t-dependence of the perturbative trajectories. The trajectories
obtained using non-perturbative propagators (solid lines) deviate substantially from the perturbative trajectories at sufficiently
small values of $-t$.

The slope of the trajectories at the origin increases as the infra-
red scale $1/a$ increases. We can see from Fig. 5.6 that a slope at

[†] The running of the coupling is assumed to stop at the infra-red scale, i.e.
$\alpha_s(\mathbf{q}^2 < 1/a^2) = \alpha_s(1/a^2)$.

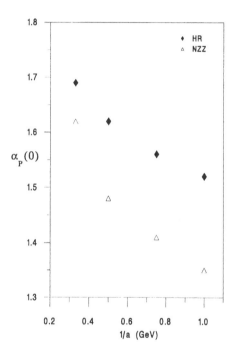

Fig. 5.4. Pomeron intercept against infra-red scale (effective gluon mass) $1/a$ for the Hancock and Ross approach (HR) and the Nikolaev, Zakharov and Zoller approach (NZZ).

the origin of 0.25 GeV as suggested by experiment would require an infra-red scale of about 0.8 GeV. On the other hand, it is quite clear from Fig. 5.5 that the trajectories are very far from linear and that the asymptotic (perturbative) solution has been reached at $-t = 1$ GeV².

The above treatment tells us that the inclusion of non-perturbative gluon propagators directly into the BFKL equation produces qualitatively desirable effects as far as the reproduction of the 'soft' Pomeron phenomenology is concerned. However, this approach is clearly far too cavalier. A somewhat more subtle procedure has been carried out by Nikolaev, Zakharov & Zoller

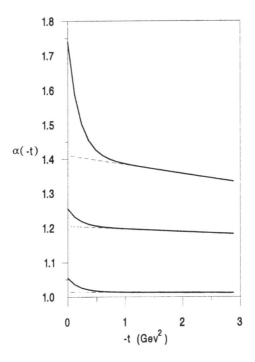

Fig. 5.5. The Pomeron and the two sub-leading trajectories as functions of momentum transfer, $-t$. The solid lines are the solutions with the inclusion of a massive gluon propagator with $1/a$ set to 0.25 GeV and the dashed lines are the results obtained using perturbative gluon propagators.

(1994a,b), using the Fock space expansion, which was briefly mentioned in the preceding chapter. In this procedure all the (BFKL) radiative corrections are incorporated in the impact factors, which are determined by considering the Fock space expansion for the wavefunction of the scattering hadron (e.g. for a meson the lowest order Fock space state is simply a quark–antiquark pair; the next is a quark–antiquark–gluon state, etc.). For each of these states a convolution is taken between the square wavefunction and the cross-section (calculated at Born level only) for the scattering pro-

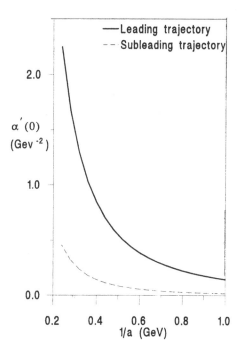

Fig. 5.6. The slopes of the Pomeron and the first sub-leading trajectories as functions of the infra-red scale (effective gluon mass) $1/a$.

cess between the Fock states. Thus, for example, the leading term for meson–meson forward scattering, where both mesons are considered to be quark–antiquark pairs, is given by

$$\mathcal{A}(s,0) = \int d^2\mathbf{b_1} d^2\mathbf{b_2} dz_1 dz_2 \, |\Psi(z_1,\mathbf{b_1})|^2 |\Psi(z_2,\mathbf{b_2})|^2 \sigma(\mathbf{b_1},\mathbf{b_2}),$$

(5.33)

where $\Psi(z_i,\mathbf{b_i})$ is the amplitude for meson i, with momentum p_i to consist of a quark–antiquark pair with longitudinal momenta $z_i p_i$, $(1-z_i)p_i$ and be separated by $\mathbf{b_i}$ in impact parameter space. The 'cross-section', $\sigma(\mathbf{b_1},\mathbf{b_2})$, is the lowest order amplitude for a process consisting of the exchange of two gluons between two quark–antiquark pairs with impact parameter separations $\mathbf{b_1}$ and $\mathbf{b_2}$ respectively. Up to a colour factor this is

$$8\alpha_s^2 \int d^2\mathbf{k} \frac{(1 - e^{i\mathbf{k}.\mathbf{b_1}})(1 - e^{i\mathbf{k}.\mathbf{b_2}})}{\mathbf{k}^4}. \qquad (5.34)$$

This procedure lends itself very easily to the incorporation of non-perturbative propagators for the long-range gluons, since the long-range gluons only appear in the cross-section. We therefore simply replace the two factors of $1/\mathbf{k}^2$ in Eq.(5.34) by $D(\mathbf{k}^2)$. Once again the simplest non-perturbative propagator is obtained by introducing an effective gluon mass $1/a$. The results for the intercept of the Pomeron obtained by Nikolaev, Zakharov & Zoller (1994a,b) are also shown in Fig. 5.4. We note that this procedure leads to a larger reduction of the intercept than the treatment by Hancock & Ross (1992). In fact, the results they obtained are lower than the lower limits given by Collins & Kwiecinski (1989) (Eq.(5.23)). However, these bounds were obtained within the context of the perturbative theory with a running coupling. Incorporation of any non-perturbative effects such as an effective gluon mass can lead to a violation of these bounds.

Collins & Landshoff (1992) have taken a somewhat different approach to the low transverse momentum behaviour of the BFKL kernel. They cut the transverse momentum off below some $\mathbf{k_0}$ in the integral for the part of the kernel which accounts for real gluon emission (the first term of Eq.(4.18)). They found that this did not shift the position of the leading singularity in the Mellin transform of the amplitude. However, they observed that there should also be an upper limit to the transverse momentum integration (from the kinematic limits) which should be of order $\sqrt{s}$. In the derivation of the BFKL equation this upper cut-off was ignored since it does not affect the leading logarithm results. Restoring the upper cut-off effectively takes into account some of the sub-leading logarithm corrections. Collins & Landshoff showed that if this upper cut-off is introduced (along with the infra-red cut-off) then the intercept of the Pomeron is reduced. McDermott, Forshaw & Ross (1995) showed that the shift downwards of the leading eigenvalue is less than 20% for $s/\mathbf{k_0}^2 > 10^4$, i.e. the leading eigenvalue is shifted to

$$\alpha_P(0) = 4\ln 2\,\bar{\alpha}_s \frac{1}{1 + \pi^2/[2 + \frac{1}{2}\ln(s/\mathbf{k_0}^2)]^2}.$$

Furthermore, in the Collins & Landshoff analysis the strong coupling was kept fixed. The running of the coupling (which is also a sub-leading logarithm effect) plays a similar role to the imposition of an upper cut-off to the transverse momentum integration. If the coupling is allowed to run then the effect of imposing an upper cut-off on the transverse momentum integral is diminished.

5.6 The heterotic Pomeron

Levin & Tan (1992) have given some consideration to the question of how one might interpolate between the 'hard' and 'soft' Pomerons. They postulate a 'heterotic Pomeron' which tends to the BFKL Pomeron when the virtuality of the external gluons is sufficiently large, and tends to the 'soft' Pomeron for near on-shell external gluons.

We introduce the impact parameter $\bar{\mathbf{b}}$, conjugate to $\mathbf{q}$, and define $\tilde{F}(s, \mathbf{k_1}, \mathbf{k_2}, \bar{\mathbf{b}})$ by

$$F(\omega, \mathbf{k_1}, \mathbf{k_2}, \mathbf{q}) = \int \frac{d^2\bar{\mathbf{b}}}{2\pi} e^{i\mathbf{q}\cdot\bar{\mathbf{b}}} \tilde{F}(s, \mathbf{k_1}, \mathbf{k_2}, \bar{\mathbf{b}}).$$

We have shown that even with the running of the coupling the dependence of the trajectories with the momentum transfer, $t = -\mathbf{q}^2$, is very small, so to a good approximation we may neglect the diffusion of the impact parameter $\bar{\mathbf{b}}$ as we go down the ladder. Therefore $\tilde{F}(s, \mathbf{k_1}, \mathbf{k_2}, \bar{\mathbf{b}})$ also (approximately) obeys the $t = 0$ BFKL equation, Eq.(4.17), which we can write (after inverting the Mellin transform) as

$$\tilde{F}(s, \mathbf{k_1}, \mathbf{k_2}, \bar{\mathbf{b}}) = \int_0^s \frac{ds'}{s'} d^2\mathbf{k}' \, \mathcal{K}_0(\mathbf{k}', \mathbf{k_1}) \tilde{F}(s', \mathbf{k}', \mathbf{k_2}, \bar{\mathbf{b}}). \quad (5.35)$$

We have explicitly written the argument of the kernel. This can be generalized to include the case where there *is* indeed substantial diffusion in the impact parameter $\bar{\mathbf{b}}$ and we can also allow for a more general energy (s) dependence. The generalized equation is then

$$\tilde{F}(s, \mathbf{k_1}, \mathbf{k_2}, \bar{\mathbf{b}}) = \int_0^s \frac{ds'}{s'} d^2\mathbf{k}' \, d^2\bar{\mathbf{b}}' \, \mathcal{K}(s/s', \mathbf{k}', \mathbf{k_1}, (\bar{\mathbf{b}} - \bar{\mathbf{b}}'))$$
$$\times \tilde{F}(s', \mathbf{k_1}, \mathbf{k_2}, \bar{\mathbf{b}}'). \quad (5.36)$$

For the 'hard' Pomeron which obtains at sufficiently large $\mathbf{k_1}, \mathbf{k'}$ the kernel is

$$\mathcal{K}(s/s', \mathbf{k'}, \mathbf{k_1}, (\bar{\mathbf{b}} - \bar{\mathbf{b}'})) = \mathcal{K}_0(\mathbf{k'} - \mathbf{k_1})\delta^2(\bar{\mathbf{b}} - \bar{\mathbf{b}'}). \qquad (5.37)$$

On the other hand, for small $\mathbf{k'}, \mathbf{k_1}$ (below $\mathbf{k_0}$) the Pomeron has a significant $\bar{\mathbf{b}}$-dependence, but does not depend much on the gluon virtuality, so we expect $\mathbf{k_1}$ to remain fixed at around $\mathbf{k_0}$. In this limit the generalized kernel has the form

$$\mathcal{K}(s/s', \mathbf{k'}, \mathbf{k_1}, (\bar{\mathbf{b}} - \bar{\mathbf{b}'})) = \delta^2(\mathbf{k'} - \mathbf{k_0}) \left(\frac{s}{s'}\right)^c B(\bar{\mathbf{b}} - \bar{\mathbf{b}'}), \quad (5.38)$$

where B is a function which vanishes as its argument becomes large, but has a non-zero width. The dynamics which determine this function are not yet understood. It cannot be derived from usual perturbation theory, but alternative techniques such as the $1/N$ expansion may shed some light on it.

The kernel Eq.(5.38) should lead to the 'soft' Pomeron, which can be described in terms of a 'ladder' in some sense (although it may not be a ladder of gluons) and as we go down the ladder we have diffusion in $\bar{\mathbf{b}}$ but not in virtuality $\mathbf{k}$.

Levin and Tan considered the case where the function $B(\bar{\mathbf{b}} - \bar{\mathbf{b}'})$ was determined by a random walk in impact parameter space as one goes down the ladder. In such a case the diffusion equation becomes

$$\frac{\partial \tilde{F}}{\partial \ln s} = c\tilde{F} + c_b \partial_{\bar{\mathbf{b}}}^2 \tilde{F}. \qquad (5.39)$$

The solution to this equation has a dependence on impact parameter, $\bar{\mathbf{b}}$, which is

$$\sim \frac{\exp(-\bar{\mathbf{b}}^2/4c_b \ln s)}{(\ln s)^{(1-c)}},$$

and which, when Fourier transformed, gives the t-dependence

$$\sim s^{c_b t}.$$

Comparing this with the experimental value for the slope of the 'soft' Pomeron trajectory we must have

$$c_b = 0.25 \text{ GeV}^{-2}.$$

The heterotic Pomeron would therefore be determined by a kernel which interpolates between the two expressions (5.37) and (5.38) as the virtuality of the external gluons varies from $\mathbf{k_1} \ll \mathbf{k_0}$

to $k_1 \gg k_0$. The soft QCD physics which gives rise to this kernel has not so far been identified. Nevertheless, the existence of a kernel, which interpolates between the 'two' Pomerons, is an intriguing possibility.

We have been discussing various attempts to explain the soft Pomeron within the context of QCD. So far none of these attempts has been particularly successful.

Nevertheless, we have the hard Pomeron which is derived from perturbative QCD using no further assumptions about the infrared behaviour. Of course, this hard Pomeron is in itself a very interesting object and it is important to put it to experimental test. In the next two chapters we shall be discussing processes such as deep inelastic scattering and large rapidity gap events in which the hard Pomeron can (at least in principle) be isolated, studied, and compared with the predictions of the 'clean' part of QCD.

5.7 Summary

• We can view the $t = 0$ BFKL equation as a diffusion equation in the transverse momentum of the emitted gluons (i.e. which make up the rungs of the ladder). Therefore, a wide range of transverse momenta contributes to the amplitude and hence it becomes necessary to consider the running of the QCD coupling.

• The BFKL equation with running coupling can be solved approximately using a technique analogous to the WKB approximation. The modified solution changes from an oscillating solution to an exponentially decaying solution above some critical transverse momentum.

• If the phase of the oscillations is fixed at some low transverse momentum by the (non-perturbative) infra-red effects of QCD, then it can only be matched for certain values of the Mellin transform variable, ω. This then leads to isolated poles for the Mellin transform of the Pomeron amplitude, as opposed to the cut obtained in the fixed coupling case.

• The infra-red phase fixing can be obtained by inserting non-perturbative gluon propagators into the BFKL equation. The intercept of the Pomeron thus obtained is reduced compared with the position of the branch point in the fixed coupling case. However, the intercept is still too large to explain, by itself, the phenomena which are so well described by the soft Pomeron.

• Non-perturbative propagators are also required to explain the quark-counting rule within the context of the two-gluon exchange model of the Pomeron. The non-perturbative propagator introduces a length scale which, if small compared to the hadron radius, will suppress quark-counting-violating contributions to the scattering amplitude in which the two gluons land on different quarks within the hadron.

• A kernel which, for large transverse gluon momenta, tends to the BFKL kernel (giving rise to diffusion in s and transverse gluon momenta but no diffusion in impact parameter) and which for small transverse gluon momenta gives rise to diffusion in s and impact parameter but *not* in gluon transverse momentum, could provide a useful interpolation between the seemingly very different 'hard' and 'soft' Pomerons.

6

Applications in deep inelastic scattering

The preceding chapters have established the theoretical framework which ought to describe the perturbative scattering of strongly interacting particles at high centre-of-mass energies (in the Regge region). In this chapter (and the next), we shall attempt to place this framework under the experimental spotlight. That is to say, we shall turn the theoretical calculations of the preceding chapters into physical cross-sections for processes which can be measured at present or future colliders.

To construct these cross-sections, we need to specify the impact factors which define the coupling of the Pomeron to the external particles. These impact factors are then convoluted with the universal BFKL amplitude, $f(\omega, \mathbf{k_1}, \mathbf{k_2}, \mathbf{q})$ (see Eq.(4.33)) in order to obtain the relevant elastic-scattering amplitude. Remember that we are using perturbation theory and so can take our result seriously only if we are sure that the typical transverse momenta are much larger than Λ_{QCD}. As we showed in Section 5.1, for $t = 0$ the largeness of the typical transverse momenta is assured provided we pick processes with impact factors which are peaked at large transverse momenta. Clearly, this is not the case for proton–proton scattering and that is why we were not surprised to find that our results were incompatible with the relatively modest rise of the p–p total cross-section with increasing s. Another way of keeping our integrals away from the infra-red region is to work at high-t but we defer this topic until the next chapter.

In this chapter we shall focus on the process of deep inelastic lepton–nucleon scattering. In the centre-of-mass frame, the incoming lepton is scattered through a large angle, radiating a highly virtual photon (γ^*) which scatters inelastically off the incoming nucleon (let us say it is a proton, p). The total cross-section for $\gamma^* p \rightarrow X$ (where X labels all possible final states) is obtained by taking the imaginary part of the elastic $\gamma^* p \to \gamma^* p$ cross-section at $t = 0$ (recall the optical theorem of Chapter 1). For high γ^*-p centre-of-mass energies we can try to use the BFKL amplitude to compute this cross-section. As we shall see, the high virtuality of the γ^* provides the large scale in the associated impact factor. Unfortunately, we also need the impact factor associated with the proton line. In this case (as is also the case in p-p scattering) we have no large scale (indeed we cannot calculate this impact factor using perturbation theory) and as such must consider the fact that our transverse momentum integrals pick up significant contributions from the infra-red region.

After illuminating the above remarks, we will consider a process which should provide a much more direct test of the purely perturbative dynamics. By picking the impact factor associated with the proton such that it describes the production of a parton of high p_T into the final state, we can sidestep the infra-red problems which plague the deep inelastic total cross-section.

In Section 6.4, we shall discuss how our approach relates to the more conventional ('Altarelli–Parisi') one. To finish the chapter we demonstrate that the assumption of multi-Regge kinematics (i.e. strong ordering of the Sudakov components) is not in general suitable as we move away from the discussion of elastic-scattering amplitudes (and hence total cross-sections).

6.1 Introduction

The basic deep inelastic amplitude for electron–proton (e-p) scattering is shown in Fig. 6.1. The incoming electron and proton four-momenta are k and p, respectively, and the virtual (space-like) photon has four-momentum q. The important kinematic invariants are

$$
\begin{aligned}
Q^2 &= -q^2 > 0, \\
s &= (p + k)^2, \\
W^2 &= (p + q)^2, \\
x &= \frac{Q^2}{2p \cdot q} \approx \frac{Q^2}{Q^2 + W^2},
\end{aligned}
$$

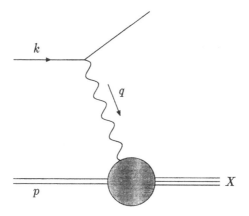

Fig. 6.1. The basic deep inelastic scattering process.

and

$$y = \frac{p \cdot q}{p \cdot k} \approx \frac{Q^2}{xs}.$$

The approximate equalities become exact in the limit of negligible lepton and proton masses. We shall subsequently assume this limit. As always, we also work in the high energy limit, i.e.

$$W^2 \gg Q^2 \gg M_p^2.$$

Since this inequality implies that $x \ll 1$ we say we are working in the low-x regime.

The total e–p cross-section can be written as a contraction of a lepton tensor (calculated purely within QED) and a hadronic tensor.[†] The hadronic tensor is then written in terms of two independent structure functions (utilizing gauge, Lorentz and time-reversal invariance, parity conservation and assuming unpolarized beams), i.e.

$$\frac{d^2\sigma}{dx\,dQ^2} = \frac{2\pi\alpha^2}{xQ^4}\left\{[1 + (1-y)^2]F_2(x, Q^2) - y^2 F_L(x, Q^2)\right\}, \quad (6.1)$$

where α is the fine structure constant. Given the assumptions, this is a completely general expression and tells us nothing about the

---

[†] For those readers unfamiliar with these details, we refer to the standard texts, e.g. Close (1979).

functional form of structure functions. However, we can re-write them in terms of the cross-sections for scattering transverse or longitudinal photons off the proton, i.e.

$$F_2(x, Q^2) = \frac{Q^2}{4\pi^2\alpha}(\sigma_T(x, Q^2) + \sigma_L(x, Q^2))$$

$$F_L(x, Q^2) = \frac{Q^2}{4\pi^2\alpha}\sigma_L(x, Q^2). \tag{6.2}$$

Our goal in the next section will be to calculate these structure functions as far as is possible (we will struggle with the proton impact factor).

6.2 The low-x structure functions

We shall obtain the structure functions by computing the imaginary part of the amplitude for elastic γ^*p scattering (for each photon polarization). In the high energy limit, for photons with polarization λ, we have (from Eq.(4.36))

$$\sigma_\lambda(x, Q^2) = \frac{\mathcal{G}}{(2\pi)^4} \int \frac{d^2\mathbf{k_1}}{\mathbf{k_1^2}} \frac{d^2\mathbf{k_2}}{\mathbf{k_2^2}} \Phi_\lambda(\mathbf{k_1})\Phi_p(\mathbf{k_2})F(x, \mathbf{k_1}, \mathbf{k_2}),$$

$$\tag{6.3}$$

where Φ_p is the proton impact factor and Φ_λ is the impact factor for a photon of polarization λ. This equation is shown graphically in Fig. 6.2 for one particular contribution to the photon impact factor (e.g. there are also contributions where the gluons couple to the different quark lines). Notice that we have shifted to a more convenient notation (with respect to Eq.(4.36)): we have suppressed all dependence on the momentum transfer $\mathbf{q}$ since it is zero and we have taken the inverse Mellin transform of the BFKL amplitude so as to obtain it as a function of $x \approx Q^2/W^2$.

We have already computed the BFKL amplitude (see Eq.(4.28)), i.e.

$$F(x, \mathbf{k_1}, \mathbf{k_2}) = \sum_{n=0}^{\infty} \int_{-\infty}^{\infty} d\nu \left(\frac{\mathbf{k_1^2}}{\mathbf{k_2^2}}\right)^{i\nu} \frac{e^{in(\theta_1 - \theta_2)}}{2\pi^2 k_1 k_2}$$

$$\times \exp\left(\bar{\alpha}_s\chi_n(\nu)\ln 1/x\right). \tag{6.4}$$

This is where the x-dependence of our final result resides and we can clearly see that the leading eigenvalue of the kernel leads to a

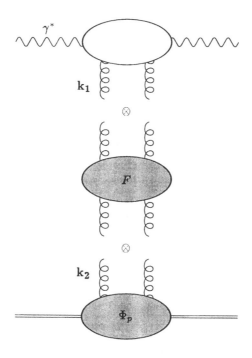

Fig. 6.2. One of the graphs contributing to the amplitude for elastic $\gamma^* p$ scattering. The '$\otimes$' symbols represent the convolution of the BFKL amplitude with the impact factors.

strong rise with decreasing x, i.e.

$$F \sim \frac{x^{-\omega_0}}{\sqrt{\ln 1/x}},$$

where ω_0 is as in Eq.(4.31), i.e. $\omega_0 = 4\bar{\alpha}_s \ln 2$.[†] This translates directly into the same rise at low x for the total deep inelastic scattering cross-section (i.e. the structure functions $F_2(x, Q^2)$ and $F_L(x, Q^2)$). We will discuss this behaviour and compare it with experimental data shortly. For now let us merely say that the data on low-x structure functions does indeed exhibit a strong rise with decreasing x. This is the first time a total cross-section has been measured which rises strongly with increasing centre-of-mass energy. Such a rise cannot be explained by the soft Pomeron

[†] Recall that $\bar{\alpha}_s = 3\alpha_s/\pi$ for the three colours of QCD.

pole which describes so well the total hadronic cross-sections and we are therefore encouraged in our attempt to use perturbative QCD.

It only remains to compute the impact factors and perform the convolution. We are unable to compute the proton impact factor using perturbation theory and so we choose to take it as a phenomenological input. We expect it to describe some primordial gluon distribution in transverse momenta which is peaked around $\sim M_p$. The photon impact factor is calculable in perturbation theory. The calculation is detailed in Appendix A to this chapter.

Using these impact factors along with Eqs.(6.3) and (6.4), we can deduce the proton structure functions up to the largely unknown proton impact factor, $\Phi_p(\mathbf{k_2})$. Before discussing this procedure further we want to deviate a little in order to introduce the concept of the gluon distribution (or density) function.

We define the **unintegrated gluon density**, $\mathcal{F}(x, \mathbf{k})$, to be that (dimensionless) 'cross-section' which would be observed if the photon impact factor (Φ_λ) were replaced by the impact factor Φ_g where

$$\Phi_g(\mathbf{k_1}) = 2\pi \mathbf{k}^4 \delta^2(\mathbf{k_1} - \mathbf{k}).$$

Thus we have the definition

$$\mathcal{F}(x, \mathbf{k}) \equiv \frac{1}{(2\pi)^3} \int \frac{d^2\mathbf{k}'}{\mathbf{k}'^2} \Phi_p(\mathbf{k}') \mathbf{k}^2 F(x, \mathbf{k}, \mathbf{k}'). \qquad (6.5)$$

The **gluon density** is then defined to be

$$G(x, Q^2) \equiv \int \frac{d^2\mathbf{k}}{\pi \mathbf{k}^2} \Theta(Q^2 - \mathbf{k}^2) \mathcal{F}(x, \mathbf{k}), \qquad (6.6)$$

where we have introduced the theta (step) function, $\Theta(Q^2 - \mathbf{k}^2)$, which is defined to equal unity when $Q^2 > \mathbf{k}^2$ and zero when $Q^2 < \mathbf{k}^2$. This definition of the gluon density will be particularly useful when we come to make our comparisons with the Altarelli-Parisi approach to the structure functions at low x. For now it merely simplifies our notation. Note that $\mathcal{F}(x, \mathbf{k})$ contains the BFKL dynamics (convoluted with the proton impact factor) but that, unlike the structure functions, it is not a physical observable.

The structure functions are thus given by

$$
\begin{aligned}
F_2(x, Q^2) &= \frac{Q^2}{4\pi^2\alpha} \int \frac{d^2\mathbf{k}}{\mathbf{k}^4} \frac{\mathcal{F}(x, \mathbf{k})}{4\pi}(\Phi_T(\mathbf{k}) + \Phi_L(\mathbf{k})) \\
&= \frac{Q^2}{4\pi^2}\alpha_s \sum_{q=1}^{n_f} e_q^2 \int \frac{d^2\mathbf{k}}{\mathbf{k}^2}\mathcal{F}(x, \mathbf{k}) \int_0^1 d\rho d\tau
\end{aligned}
$$

$$
\times \; \frac{1 - 2\rho(1-\rho) - 2\tau(1-\tau) + 12\rho(1-\rho)\tau(1-\tau)}{Q^2\rho(1-\rho) + \mathbf{k}^2\tau(1-\tau)} \tag{6.7}
$$

and

$$
\begin{aligned}
F_L(x, Q^2) &= \frac{Q^2}{4\pi^2\alpha} \int \frac{d^2\mathbf{k}}{\mathbf{k}^4} \frac{\mathcal{F}(x, \mathbf{k})}{4\pi}\Phi_L(\mathbf{k}) \\
&= \frac{2Q^2}{\pi^2}\alpha_s \sum_{q=1}^{n_f} e_q^2 \int \frac{d^2\mathbf{k}}{\mathbf{k}^2}\mathcal{F}(x, \mathbf{k}) \int_0^1 d\rho d\tau
\end{aligned}
$$

$$
\times \; \frac{\rho(1-\rho)\tau(1-\tau)}{Q^2\rho(1-\rho) + \mathbf{k}^2\tau(1-\tau)}. \tag{6.8}
$$

We have multiplied by the factor $T(F) = \frac{1}{2}$, to account for the colour factor associated with the upper quark loop. The impact factors, Φ_T and Φ_L, are for scattering off transverse and longitudinal photons, respectively, and their calculation is detailed in Appendix A to this chapter.

Note that our definition of $G(x, Q^2)$ is such that, in the limit

$$
\mathbf{k}^2 \ll Q^2
$$

(which is equivalent to taking the leading log Q^2 approximation to the BFKL equation and is usually referred to as the **double leading log approximation**), we can write

$$
\begin{aligned}
\frac{\partial F_2(x, Q^2)}{\partial \ln Q^2} &= 2\sum_{q=1}^{n_f} e_q^2 \frac{\bar{\alpha}_s}{6} \int_0^1 d\tau P_{qg}(\tau) G(x, Q^2) \\
&= \sum_{q=1}^{n_f} e_q^2 \frac{\bar{\alpha}_s}{9} G(x, Q^2), \tag{6.9}
\end{aligned}
$$

where $P_{qg}(\tau) = \frac{1}{2}(\tau^2 + (1-\tau)^2)$ is the usual Altarelli–Parisi splitting function. The key to obtaining this expression is to notice that, after differentiating Eq.(6.7) with respect to $\ln Q^2$, the dominant contribution to the ρ integral is from the end-points where ρ is within $\tau(1-\tau)\mathbf{k}^2/Q^2$ of either 0 or 1 (modulo terms which are suppressed by $\sim k^2/Q^2$). Later we shall discuss the connection between the Altarelli–Parisi and the BFKL approaches in more detail.

Let us conclude this section by introducing a toy model for the proton impact factor. We can then compute the low-x structure functions. We will try an impact factor of the form

$$\Phi_p(\mathbf{k}) \sim \left(\frac{\mathbf{k}^2}{\mathbf{k}^2 + \mu^2}\right)^\delta, \tag{6.10}$$

where μ is a scale which is typical of the non-perturbative dynamics and δ is essentially unknown (except that we know that $\delta > \frac{1}{2}$ in order that the $\mathbf{k}'$ integral is finite). The $\mathbf{k}'$ integral can now be performed since

$$\int \frac{d^2\mathbf{k}'}{\mathbf{k}'^2} \left(\frac{\mathbf{k}'^2}{\mathbf{k}'^2 + \mu^2}\right)^\delta (\mathbf{k}'^2)^{-1/2-i\nu}$$

$$= \pi(\mu^2)^{-1/2-i\nu} \frac{\Gamma(\delta - 1/2 - i\nu)\Gamma(1/2 + i\nu)}{\Gamma(\delta)}. \tag{6.11}$$

Thus, in the $n = 0$ limit, which selects the leading eigenvalue of the kernel (the angular integrals are then trivial), we have

$$\frac{\mathcal{F}(x, \mathbf{k})}{\mathbf{k}^2} = \frac{\mathcal{N}_g}{2\pi} \int_{-\infty}^{\infty} \frac{d\nu}{\sqrt{\mu^2 \mathbf{k}^2}} \left(\frac{\mathbf{k}^2}{\mu^2}\right)^{i\nu} \exp\left(\bar{\alpha}_s \chi_0(\nu) \ln 1/x\right)$$

$$\times \frac{\Gamma(\delta - 1/2 - i\nu)\Gamma(1/2 + i\nu)}{\Gamma(\delta)}. \tag{6.12}$$

The constant $\mathcal{N}_g$ contains the unknown normalization of the proton impact factor as well as the colour factor and factors of π. Substituting this into the expression for $F_2(x, Q^2)$, we can also perform the $\mathbf{k}$ integral, using

$$\int_0^\infty \frac{(\mathbf{k}^2)^{-1/2+i\nu} d\mathbf{k}^2}{Q^2 \rho(1 - \rho) + \mathbf{k}^2 \tau(1 - \tau)} = \frac{\pi[\tau(1 - \tau)]^{-1}}{\cosh \pi\nu}$$

$$\times \left[\frac{Q^2 \rho(1 - \rho)}{\tau(1 - \tau)}\right]^{-1/2+i\nu}. \tag{6.13}$$

Putting all this together we obtain the result that

$$F_2(x, Q^2) = \left(\frac{Q^2}{\mu^2}\right)^{1/2} \frac{\mathcal{N}_g \alpha_s}{8\pi} \sum e_q^2 \int_0^1 d\rho \int_0^1 d\tau \int_{-\infty}^{\infty} d\nu$$

$$\times \frac{1}{\cosh\pi\nu} \left(\frac{Q^2}{\mu^2}\right)^{i\nu} \exp\left(\bar{\alpha}_s \chi_0(\nu)\ln 1/x\right)$$

$$\times \left[1 - 2\rho(1 - \rho) - 2\tau(1 - \tau) + 12\tau\rho(1 - \tau)(1 - \rho)\right]$$

$$\times \left[\frac{\rho(1 - \rho)}{\tau(1 - \tau)}\right]^{-1/2+i\nu} \frac{\Gamma(\delta - 1/2 - i\nu)\Gamma(1/2 + i\nu)}{\tau(1 - \tau)\Gamma(\delta)}. \tag{6.14}$$

We can perform the ν-integral by expanding about the saddle point[†] at $\nu \approx 0$. This has the effect of decoupling all the δ-dependence and renders the ρ- and τ-integrals purely numerical. Thus we can factorize them into some new (and unknown) constant, $\mathcal{N}_2$. Our final result for $F_2(x, Q^2)$ is therefore

$$F_2(x, Q^2) \approx \mathcal{N}_2 \bar{\alpha}_s \sum e_q^2 \left(\frac{Q^2}{\mu^2}\right)^{1/2} \frac{e^{\omega_0 \ln 1/x}}{\sqrt{\bar{\alpha}_s \ln 1/x}}$$

$$\times \exp\left(-\frac{\ln^2(Q^2/\mu^2)}{56\bar{\alpha}_s \zeta(3)\ln 1/x}\right). \tag{6.15}$$

In Fig. 6.3, we show a sample of the HERA data collected from e-p collisions at a centre-of-mass energy $\sqrt{s} \approx 300$ GeV. The curves arise from Eq.(6.15). What exactly did we do with Eq.(6.15) in order to produce these curves? The answer to this question highlights the difficulty in making firm predictions for the structure function $F_2(x, Q^2)$. The normalization, $\mathcal{N}_2$, is unknown (it depends on non-perturbative physics through the normalization of the proton impact factor and our *ansatz* for its shape, i.e. δ in our toy model) – so we need to fit it to the data. The scale μ^2 is again of non-perturbative origin. Additionally, we do not know the appropriate value of α_s to take nor do we know the appropriate scale to define the logarithms of energy (i.e. do we take $\ln 1/x$ or $\ln 1/2x$, etc?). These are both problems which originate because we only summed the leading logarithms in energy and can only be improved by going beyond this approximation. For

---

[†] A brief introduction to the saddle point method is given in Appendix B to this chapter.

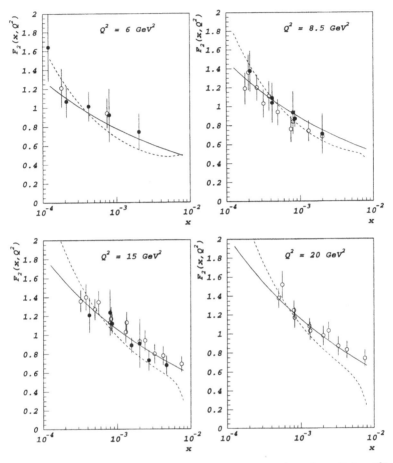

Fig. 6.3. Data on the deep inelastic structure function, $F_2(x, Q^2)$, collected by the H1 (open circles) (Ahmed *et al.* (1995)) and Zeus (full circles) (Derrick *et al.* (1995)) collaborations at the HERA e–p collider.

the solid line we chose $\alpha_s = 0.1$ and introduced a parameter, x_0, such that all occurrences of $\ln 1/x$ are replaced by $\ln x_0/x$. The parameters x_0, $\mathcal{N}_2$ and μ^2 were then fitted to the data. The best fit values were $x_0 = 0.6$, $\mathcal{N}_2 = 0.30$ and $\mu^2 = 0.31$ GeV^2. The ambiguity in x_0 is equivalent to saying that we really cannot answer the question "how low in x must we be to observe the dynamics associated with the leading logarithm summation?", since this region is defined to be that for which $x \ll x_0$. Correspondingly, we fitted only to data which satisfy this criterion, i.e. we fitted to data with $x < 10^{-2}$. Even so, with this four parameter fit, we are able to obtain good agreement with the low-x data. The dashed line illustrates the strong sensitivity to our choice of α_s. It is produced with $\alpha_s = 0.2$. The other parameters are correspondingly re-fitted, i.e. $x_0 = 0.01$, $\mu^2 = 2$ GeV^2 and $\mathcal{N}_2 = 0.38$, and the fit is to all those data with $x < 10^{-3}$. Since $\omega_0 \propto \alpha_s$ drives the low-x rise we should not be surprised to see a much steeper behaviour with $\alpha_s = 0.2$.

We took a model for the proton impact factor (Eq.(6.10)) which (after dividing by $\mathbf{k}^2$) is peaked in the region of low $\mathbf{k}^2$. In light of the discussion in Section 5.1 of the preceding chapter, we should question the validity of our calculation. The diffusion 'cigar' is tilted (one end fixed by $\sim Q^2$ and the other by $\sim \mu^2$) and as a result there is the danger that contributions from the infra-red region could possibly be large. At the end of this chapter, when we make the connection with the Altarelli–Parisi approach, we shall show that (neglecting terms suppressed by powers of $\sim \mu^2/Q^2$) the largely unknown infra-red physics factorizes from the known perturbative physics. This is good news since it means that we can make meaningful perturbative calculations. We defer further studies on the total inelastic cross-section to the end of this chapter and turn to a process which avoids many of the problems associated with the unknown infra-red effects we have just been discussing.

6.3 Associated jet production

Although the total deep inelastic cross-section is relatively straightforward to measure, the work of the last section has taught us that the non-perturbative behaviour of the proton impact factor spoils a clean perturbative analysis. Our problem was with the fact that the proton impact factor introduced unknown non-perturbative effects into our calculation. We modelled them at the price of introducing unknown parameters μ^2 and δ. If we could replace this impact factor with one which is peaked at a much larger scale then we eliminate most of our difficulties (we always need to worry about the effects of diffusion if the centre-of-mass energy is

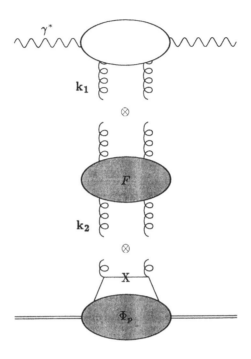

Fig. 6.4. One of the amplitudes relevant for associated jet production. The lower quark line could also be a gluon line and the 'X'
denotes a high-p_T parton.

too big). Mueller (1991) appreciated that it is possible to do this
by insisting on the production of a high-p_T parton which emerges
at a small angle relative to the direction of the incoming proton
(e.g. in the γ^*p CM frame). In Fig. 6.4 we show the relevant amplitude. The cross on the lower parton line indicates that it has
a high transverse momentum. The initial studies into associated
jet production can be found in the papers by Kwiecinski, Martin
& Sutton (1992), Bartels, De Roeck & Loewe (1992) and Tang
(1992).

We say that the high-p_T parton is in the forward direction, i.e. it
carries a not-too-small fraction, x_j, of the incoming proton energy
(relative to its transverse momentum). Of course, experimental-
ists do not measure this parton. They observe the jet of hadrons
which it produces. In this limit, we may neglect the transverse

momentum of the parton from the proton. Consequently, if the associated jet is a quark of flavour i, the proton impact factor, Φ_p^i, becomes

$$\Phi_p^i(\mathbf{k_2}) = 8\pi^2\alpha_s q_i(x_j, \mathbf{k_2^2}), \qquad (6.16)$$

i.e. the quark impact factor of Eq.(4.37) multiplied by the quark number density. The colour factor for this process is now $\mathcal{G} = NG_0^{(1)}$ (since we do not average over the colour of the quark loop in the photon impact factor). Putting the number of colours equal to 3 (as we shall do subsequently in this chapter), $\mathcal{G} = \frac{2}{3}$. Similarly,

$$\Phi_p^g(\mathbf{k_2}) = 8\pi^2\alpha_s g(x_j, \mathbf{k_2^2}), \qquad (6.17)$$

for a gluon jet and the colour factor is now $\frac{1}{8}f_{abc}f_{abd}\mathrm{Tr}(T_c T_d) = \frac{3}{2}$, i.e. it is $C_2(A)/C_2(F) = \frac{9}{4}$ larger than the quark colour factor.

By observing the associated jet, we fix $\mathbf{k_2} = \mathbf{k_j}$ (assuming the incoming parton to be collinear with the proton). Thus, for the differential structure functions we can write:

$$x_j\mathbf{k_j^2}\frac{\partial^2 F_\lambda(x, Q^2; x_j, \mathbf{k_j^2})}{\partial x_j \partial \mathbf{k_j^2}} = \frac{Q^2}{4\pi^2\alpha}8\pi^2\alpha_s\frac{3}{2}\frac{\pi}{(2\pi)^4}$$

$$\times \int \frac{d^2\mathbf{k}}{\mathbf{k}^2}\Phi_\lambda(\mathbf{k})F(x/x_j, \mathbf{k}, \mathbf{k_j})\left[G(x_j, \mathbf{k_j^2}) + \frac{4}{9}\Sigma(x_j, \mathbf{k_j^2})\right], \quad (6.18)$$

where

$$G(x, Q^2) \equiv xg(x, Q^2) \quad \text{and}$$

$$\Sigma(x, Q^2) \equiv \sum_{i=1}^{n_f}[xq_i(x, Q^2) + x\bar{q}_i(x, Q^2)] \qquad (6.19)$$

are the momentum distribution functions. Notice that we have taken x/x_j in the arguments of F. Since $x_j \gg x$ this is, strictly speaking, sub-leading. However our choice reflects the fact that the γ^*–parton sub-energy is $\approx Q^2(x_j/x)$.

By looking at forward jets, i.e. $x_j \sim 1$ and $\mathbf{k_j^2} \sim Q^2$, we are focusing on a region where the parton densities, $G(x_j, \mathbf{k_j^2})$ and $\Sigma(x_j, \mathbf{k_j^2})$, are experimentally well measured. In this way we have avoided any need to invoke non-perturbative effects directly since they have been implicitly factorized into the parton density functions.

Let us focus on the scattering of transverse photons, the impact factor is given in Eq.(A.6.21) of Appendix A to this chapter. Again we can perform the k-integral (using Eq.(6.13)). Keeping only the $n = 0$ term in the BFKL expansion it then follows that

$$x_j \mathbf{k_j^2} \frac{\partial^2 F_T(x, Q^2; x_j, \mathbf{k_j^2})}{\partial x_j \partial \mathbf{k_j^2}} = \frac{\pi \bar{\alpha}_s^2}{3} \left[G(x_j, \mathbf{k_j^2}) + \frac{4}{9} \Sigma(x_j, \mathbf{k_j^2}) \right]$$

$$\times \sum_{q=1}^{n_f} e_q^2 \int_{-\infty}^{\infty} \frac{d\nu}{8\pi} \frac{1}{\cosh \pi \nu} \int \frac{d\rho d\tau}{\tau(1 - \tau)}$$

$$\times [\tau^2 + (1 - \tau)^2][\rho^2 + (1 - \rho)^2] \left[\frac{\rho(1 - \rho)}{\tau(1 - \tau)} \right]^{-1/2 + i\nu}$$

$$\times \left(\frac{Q^2}{\mathbf{k_j^2}} \right)^{1/2 + i\nu} \exp\left(\bar{\alpha}_s \chi_0(\nu) \ln(x_j/x) \right). \tag{6.20}$$

Performing the ν-integral by the saddle point method about $\nu = 0$ and using

$$\int_0^1 d\rho \frac{\rho^2 + (1 - \rho)^2}{\sqrt{\rho(1 - \rho)}} = \frac{3\pi}{4}, \tag{6.21}$$

gives

$$x_j \mathbf{k_j^2} \frac{\partial^2 F_T(x, Q^2; x_j, \mathbf{k_j^2})}{\partial x_j \partial \mathbf{k_j^2}} \approx \left[G(x_j, \mathbf{k_j^2}) + \frac{4}{9} \Sigma(x_j, \mathbf{k_j^2}) \right]$$

$$\times \bar{\alpha}_s A_T \sum_{q=1}^{n_f} e_q^2 \left(\frac{\pi \bar{\alpha}_s}{126 \zeta(3) \ln(x_j/x)} \right)^{1/2} \left(\frac{Q^2}{\mathbf{k_j^2}} \right)^{1/2}$$

$$\times \left(\frac{x_j}{x} \right)^{\omega_0} \exp\left(-\frac{\ln^2(Q^2/\mathbf{k_j^2})}{56 \zeta(3) \bar{\alpha}_s \ln(x_j/x)} \right), \tag{6.22}$$

where $A_T = 9\pi^2/128$.

In the case of longitudinal photons the calculation proceeds along similar lines and it is straightforward to show that the form is just as for transverse photons, but with A_T replaced by $A_L = \pi^2/64$. Since $F_2(x, Q^2) = F_L(x, Q^2) + F_T(x, Q^2)$, we have $A_2 = 11\pi^2/128$.

We have suggested that $\mathbf{k_j^2}$ should be chosen to be $\sim Q^2$. This has the clear benefit (provided $Q^2 \gg \Lambda_{QCD}^2$) of ensuring that there is no danger of diffusion effects forcing the loop integrals (which are implicit in the BFKL amplitude, $F(x_j/x, k^2, \mathbf{k_j^2})$) to pick up large contributions from the infra-red region. In the language of Section 5.1, the axis of the diffusion 'cigar' is horizontal and well above the dangerous small τ' region. However, since the effect of diffusion increases with increasing x_j/x there is a limit to how low in x we can go before non-perturbative effects start to become important in the calculation of the BFKL amplitude, i.e. the diffusion 'cigar' becomes too fat. For a more detailed discussion on the effects of diffusion we refer to the paper by Bartels & Lotter (1993).

One other advantage of choosing $\mathbf{k}_{\mathbf{j}}^2 \sim Q^2$ arises once we appreciate that the $\ln 1/x$ terms that we have been concentrating so hard on summing up are not the only logarithms that can be large. There are also $\ln Q^2$ terms which can compete in the deep inelastic regime. The Altarelli–Parisi equations tell us how to sum the $\ln Q^2$ terms which occur in the perturbative expansion. These terms compete with the $\ln 1/x$ terms and ideally we should look at both series in a complete treatment. However, by picking $\mathbf{k}_{\mathbf{j}}^2 \sim Q^2$ we ensure that there are no large logarithms in Q^2 in the BFKL amplitude (only $\ln Q^2/\mathbf{k}_{\mathbf{j}}^2$ terms appear). The summation of the large $\ln \mathbf{k}_{\mathbf{j}}^2$ logarithms is implicit in the parton densities, $G(x_j, \mathbf{k}_{\mathbf{j}}^2)$ and $\Sigma(x_j, \mathbf{k}_{\mathbf{j}}^2)$, which, as we have said, we are able to read off from experiment.

Let us conclude our discussion of the associated jet process with a short study of the feasibility of its experimental detection. The main difficulty associated with insisting on seeing a forward jet arises precisely because the jet is *forward*. There is a limit to how forward the jet can go, since we need it to appear in the detector (i.e. not vanish down the beam-pipe). Also, since there are other particles heading down the forward beam-pipe (from the break up of the proton) our jet had better be sufficiently well collimated and isolated. If Θ is the minimum angle at which the parton can emerge (in the lab frame) so that the associated jet is observable, i.e. it appears as a discernible jet in the detectors, then it follows that

$$\frac{\sqrt{\mathbf{k}_{\mathbf{j}}^2}}{x_j p} > \tan\Theta.$$

In addition, the high energy limit demands that $x/x_j \ll 1$. So we

Applications in deep inelastic scattering

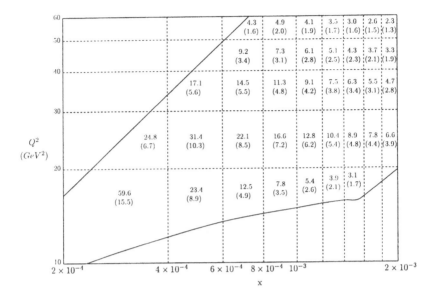

Fig. 6.5. Cross-section for the associated jet process. The cross-sections are in pb.

have competing constraints which lead to important cuts on the allowed phase space. These problems are inherently due to the configuration of the experiment (and the existence of a proton remnant) and cannot be circumvented simply by increasing the proton beam momentum (since this makes the jet more forward) nor by increasing the electron beam momentum (since we need to detect the scattered electron). In Fig. 6.5, cross-sections for the associated jet process are shown in different $x - Q^2$ bins. The HERA collider is used to define the proton (820 GeV) and electron (30 GeV) lab frame energies and the typical acceptance cuts. The boundary lines are due to the cuts on the angle of the scattered electron and the cross-sections are computed with the following kinematical constraints: $\Theta = 5°$, $x_j > 0.05$, $x_j/x > 10$, $Q^2/2 < \mathbf{k}_j^2 < 2Q^2$. The numbers in parentheses are the cross-sections calculated without BFKL corrections, i.e. $F(s, \mathbf{k}_1, \mathbf{k}_2, 0) = \delta^2(\mathbf{k}_1 - \mathbf{k}_2)$. We took these results from the paper by Martin, Kwiecinski & Sutton (1992).

6.4 The Altarelli–Parisi approach

In most introductory text books on QCD one will find a discussion of the quark–parton model of the proton. The impulse approximation allows us to consider (in the infinite momentum frame) the proton as a system of partons whose transverse motion is frozen over the time scales typical of the interaction with the off-shell photon. The partons are point-like and so we are led to the concept of Bjorken scaling, i.e. the Q^2-independence of the deep inelastic structure function, $F_2(x, Q^2)$, and the vanishing of the longitudinal cross-section (the Callan-Gross relation) in the $Q^2 \to \infty$ limit.

The experimental data is consistent with this picture to a fair approximation but scaling violations are seen. These violations can be accounted for within the framework of QCD perturbation theory using the so-called Altarelli–Parisi equations. Subsequently, we will refer to these equations more correctly as the **DGLAP equations**, after Dokshitzer (1977), Gribov & Lipatov (1972) and Altarelli & Parisi (1977). The DGLAP equations rely on the notion of proton quark, $q_i(x, Q^2)$, and gluon, $g(x, Q^2)$, density functions which specify the number density of partons within the proton. For example, the first moment of the quark density (summed over all quark flavours) is to be interpreted as the fraction of the proton's momentum carried by the quarks, i.e.

$$\int_0^1 dx[\Sigma(x, Q^2) + G(x, Q^2)] = 1,$$

where $G(x, Q^2)$ and $\Sigma(x, Q^2)$ are defined as in Eq.(6.19).

Let us recall[†] the DGLAP equations:

$$\frac{\partial}{\partial \ln Q^2} \left(\begin{array}{c} \Sigma(x, Q^2) \\ G(x, Q^2) \end{array} \right)$$
$$= \frac{\alpha_s}{2\pi} \int_x^1 dz \left(\begin{array}{cc} P_{qq}(z) & 2n_f P_{qg}(z) \\ P_{gq}(z) & P_{gg}(z) \end{array} \right) \left(\begin{array}{c} \Sigma(x/z, Q^2) \\ G(x/z, Q^2) \end{array} \right). \quad (6.23)$$

These equations are easiest to solve in moment space, i.e. we take the Mellin transforms of the parton densities using the fact that

---

[†] For an introduction to the DGLAP formalism, we refer to the standard texts, e.g. Field (1989), Halzen & Martin (1984), Greiner & Schäfer (1994), Roberts (1990).

for Q^2 fixed, $s \propto 1/x$. Accordingly we define

$$\begin{pmatrix} \Sigma_N(Q^2) \\ G_N(Q^2) \end{pmatrix} = \int_0^1 dx \, x^{N-1} \begin{pmatrix} \Sigma(x, Q^2) \\ G(x, Q^2) \end{pmatrix}, \qquad (6.24)$$

$$\gamma_{ij}^N = \frac{\alpha_s}{2\pi} \int_0^1 dz \, z^N P_{ij}(z).$$

The moment index, N, is not to be confused with the number of colours! Although in preceding chapters we labelled the moment index ω, in this chapter we adopt the notation which is most common in the literature when discussing the deep inelastic structure functions. The DGLAP equations now reduce to a pair of simultaneous equations:

$$\frac{\partial}{\partial \ln Q^2} \begin{pmatrix} \Sigma_N(Q^2) \\ G_N(Q^2) \end{pmatrix} = \begin{pmatrix} \gamma_{qq}^N & 2n_f \gamma_{qg}^N \\ \gamma_{gq}^N & \gamma_{gg}^N \end{pmatrix} \begin{pmatrix} \Sigma_N(Q^2) \\ G_N(Q^2) \end{pmatrix}. \qquad (6.25)$$

This is easy to see after inserting a Dirac delta function to write

$$\int_x^1 dz \, P_{ij}(z) f(x/z) = \int_0^1 dz \int_0^1 dy \, z P_{ij}(z) f(y) \delta(x - yz).$$

The solution is now straightforward to obtain. The matrix, γ_{ij}^N, is called the **anomalous dimension** matrix (for reasons that will become clear) and is calculable in perturbation theory.

In terms of the parton densities the deep inelastic structure functions can be written in the simple form:

$$\begin{aligned} F_{\lambda,N}(Q^2) &= \sum_i e_i^2 C_{i,N}^{(\lambda)}(Q^2/\mu_F^2, \alpha_s(\mu_F^2)) Q_{i,N}(\mu_F^2) \\ &\quad + C_{g,N}^{(\lambda)}(Q^2/\mu_F^2, \alpha_s(\mu_F^2)) G_N(\mu_F^2), \end{aligned} \qquad (6.26)$$

where $Q_{i,N}(\mu_F^2)$ is the Nth moment of $x q_i(x, \mu_F^2)$. The coefficient functions, $C_{(i,g),N}^{(\lambda)}$, are computable in perturbation theory, i.e. all the long distance physics factorizes into the parton densities. The factorization scale, μ_F^2, is arbitrary and the final result does not depend upon it to the given order (in α_s) of the calculation, i.e. the μ_F^2-dependence is sub-leading in α_s. It is usually best to take $\mu_F^2 = Q^2$, so that terms $\sim \alpha_s(\mu_F^2) \ln Q^2/\mu_F^2$ do not appear in the coefficient functions but are absorbed into the parton densities. This factorization of long- and short-distance physics is a fundamental and very important ingredient of QCD. The parton density functions are universal, e.g. the cross-section for Drell–Yan production of muon pairs in hadron–hadron colliders can be written as a product of two parton density functions (one for each hadron) and a hard sub-process cross-section (i.e. $q\bar{q} \to \mu^{+}\mu^{-}$). The parton densities are just those which appear in the deep inelastic structure functions. Factorization is proven only for the 'leading twist' component of the matrix elements (we refer to the review by Collins, Soper & Sterman (1989) and references therein for further details). In the case of the deep inelastic structure functions this means that factorization applies to the cross-section after we have thrown away all terms which vanish as $Q^2 \to \infty$.

By way of illustration let us consider the Nth moment of the structure function $F_{2,N}(Q^2)$. In the lowest order of perturbation theory, the gluon coefficient function vanishes (since the gluon carries no electromagnetic charge) and the quark coefficient functions are simply unity, i.e.

$$F_2(x, Q^2) = \sum_{i=1}^{n_f} e_i^2 x [q_i(x, Q^2) + \bar{q}_i(x, Q^2)]. \qquad (6.27)$$

The quark densities are obtained by solving the DGLAP equations with the lowest order splitting functions, i.e.

$$\gamma_{qq}^N = \frac{\bar{\alpha}_s}{9} \left(\frac{2}{(N+1)(N+2)} - 1 - 4 \sum_{j=2}^{N+1} \frac{1}{j} \right),$$

$$\gamma_{qg}^N = \frac{\bar{\alpha}_s}{12} \frac{N^2 + 3N + 4}{(N+1)(N+2)(N+3)},$$

$$\gamma_{gq}^N = \frac{2\bar{\alpha}_s}{9} \frac{N^2 + 3N + 4}{N(N+1)(N+2)},$$

$$\gamma_{gg}^N = -\frac{\bar{\alpha}_s}{2} \left(\frac{(N-1)(N+2)}{N(N+1)} + \frac{(N-1)(N+6)}{6(N+2)(N+3)} \right.$$

$$\left. + \sum_{j=3}^{N+1} \frac{2}{j} + \frac{1}{9} n_f \right). \qquad (6.28)$$

To solve Eq.(6.25), we need to fix the boundary conditions by specifying the parton densities at some scale, μ^2. The solution then gives the parton densities at all other scales. The parton densities which are obtained as the solution to Eq.(6.25), which is obtained using the (lowest order) anomalous dimensions of Eq.(6.28), include all perturbative corrections to the inputs $(q(x, \mu^2)$ and

$g(x, \mu^2))$ which are $\sim (\alpha_s(\mu_F^2)\ln \mu_F^2/\mu^2)^n$, i.e. the corrections are computed to leading $\ln Q^2$ accuracy. At the next order (i.e. including those terms which are $\sim \alpha_s^2$ in the anomalous dimension matrix and $\sim \alpha_s$ in the coefficient functions) the DGLAP equations sum the next-to-leading logarithms, $\sim \alpha_s(\mu_F^2)(\alpha_s(\mu_F^2)\ln \mu_F^2/\mu^2)^n$.

Provided we take Q^2 large enough (so that the leading twist terms dominate) then we expect to be able to factorize the non-perturbative behaviour of the BFKL amplitude and to re-write it in a way which is consistent with the low-x limit of the DGLAP formalism. Let us now investigate how this comes about.

We start with the unintegrated gluon density of Eq.(6.5). It will be useful to introduce the variables γ and N which are the Mellin conjugates to $\mathbf{k}^2$ and x, respectively, i.e.

$$\mathcal{F}_N(\mathbf{k}) = \int_0^1 dx\, x^{N-1} \mathcal{F}(x, \mathbf{k}),$$

$$\tilde{\mathcal{F}}_N(\gamma) = \int_1^\infty d\left(\frac{\mathbf{k}^2}{\mu^2}\right) \left(\frac{\mathbf{k}^2}{\mu^2}\right)^{-\gamma-1} \mathcal{F}_N(\mathbf{k}). \qquad (6.29)$$

Equation (6.29) can be inverted using

$$\mathcal{F}_N(\mathbf{k}) = \int_{1/2-i\infty}^{1/2+i\infty} \frac{d\gamma}{2\pi i} \left(\frac{\mathbf{k}^2}{\mu^2}\right)^\gamma \tilde{\mathcal{F}}_N(\gamma). \qquad (6.30)$$

With these definitions, we thus have

$$\mathcal{F}_N(\mathbf{k}) = \frac{1}{(2\pi)^3} \int \frac{d^2\mathbf{k}'}{\pi \mathbf{k}'^2} \Phi_p(\mathbf{k}') \int_{1/2-i\infty}^{1/2+i\infty} \frac{d\gamma}{2\pi i}$$
$$\times \left(\frac{\mathbf{k}^2}{\mathbf{k}'^2}\right)^\gamma \frac{1}{N - \bar{\alpha}_s \chi(\gamma)}. \qquad (6.31)$$

As usual, we have kept only the $n = 0$ term. Also, we substituted γ for $1/2 + i\nu$ and defined $\chi(\gamma) \equiv \chi_0(\nu)$. Using Eq.(4.27) this means that

$$\chi(\gamma) = -2\gamma_E - \psi(\gamma) - \psi(1 - \gamma).$$

The $\mathbf{k}'$-integral can be performed (it simply takes the Mellin transform of the proton impact factor). Thus,

$$\mathcal{F}_N(\mathbf{k}) = \frac{1}{(2\pi)^3} \int_{1/2-i\infty}^{1/2+i\infty} \frac{d\gamma}{2\pi i} \tilde{\Phi}_p(\gamma, \mu) \left(\frac{\mathbf{k}^2}{\mu^2}\right)^\gamma \frac{1}{N - \bar{\alpha}_s \chi(\gamma)}. \qquad (6.32)$$

Equivalently, we may write

$$\tilde{\mathcal{F}}_N(\gamma) = \tilde{\mathcal{F}}_N^0(\gamma, \mu) \frac{N}{N - \bar{\alpha}_s \chi(\gamma)}, \qquad (6.33)$$

where

$$\tilde{\mathcal{F}}_N^0(\gamma, \mu) = \frac{1}{(2\pi)^3} \frac{\Phi_p(\gamma, \mu)}{N} \qquad (6.34)$$

is the (double Mellin transform of the) unintegrated gluon density in the absence of any QCD corrections.

So far we have merely repeated the work discussed earlier in this chapter, albeit in a different notation. The moments of the structure functions $F_{i,N}(Q^2)$ also take on simple forms in this notation. Concentrating on $F_{2,N}(Q^2)$ we have

$$F_{2,N}(Q^2) = \frac{\pi \bar{\alpha}_s}{12} \sum_{q=1}^{n_f} e_q^2 \int \frac{d\gamma}{2\pi i} \tilde{\mathcal{F}}_N(\gamma) \int_0^1 \frac{d\rho \, d\tau}{\sin \pi \gamma} \left(\frac{Q^2}{\mu^2} \right)^\gamma \quad (6.35)$$

$$\times \left[\frac{\rho(1-\rho)}{\tau(1-\tau)} \right]^\gamma \frac{1 - 2\rho(1-\rho) - 2\tau(1-\tau) + 12\rho(1-\rho)\tau(1-\tau)}{\rho(1-\rho)}.$$

To obtain this result, we needed to use Eq.(6.13) to perform the $\mathbf{k}'^2$ integral. The ρ and τ integrals can also be done (they are standard integrals) and yield

$$F_{2,N}(Q^2) = \int \frac{d\gamma}{2\pi i} \frac{h_{2,N}(\gamma)}{\gamma^2} \tilde{\mathcal{F}}_N(\gamma) \left(\frac{Q^2}{\mu^2} \right)^\gamma, \qquad (6.36)$$

where

$$h_{2,N}(\gamma) = \frac{\pi \bar{\alpha}_s}{24} \sum_{q=1}^{n_f} e_q^2 \frac{\pi \gamma}{\sin \pi \gamma} \frac{\Gamma(1+\gamma)\Gamma(1-\gamma)}{\Gamma(3/2+\gamma)\Gamma(3/2-\gamma)} \frac{2 + 3\gamma - 3\gamma^2}{3 - 2\gamma}.$$

We are then left with only the integral over γ to perform.

The leading twist behaviour is specified by those contributions which do not vanish as $Q^2/\mu^2 \to \infty$. Since $Q^2 > \mu^2$, we need to close the γ-plane contour in the left half plane. There are two poles which lead to finite contributions as $Q^2/\mu^2 \to \infty$:
(a) the pole $\bar{\gamma} > 0$ which satisfies $N = \bar{\alpha}_s \chi(\bar{\gamma})$;
(b) the pole at $\gamma = 0$ (it is only a simple pole since $\chi(\gamma) \sim 1/\gamma$ as $\gamma - 0$).
All other poles (which occur for negative integer values of γ) lead to contributions which are suppressed by powers of μ^2/Q^2. The pole at $\gamma = 0$ leads to a scaling (i.e. Q^2-independent) contribution

and can be absorbed into the input quark density. Thus to reveal the predicted scaling violations we focus on $\partial F_{2,N}(Q^2)/\partial \ln Q^2$.

$$\frac{\partial F_{2,N}(Q^2)}{\partial \ln Q^2} = h_{2,N}(\bar{\gamma}) R_N \mathcal{F}_N^0(\bar{\gamma}) \left(\frac{Q^2}{\mu^2}\right)^{\bar{\gamma}}, \qquad (6.37)$$

where

$$R_N = \frac{1}{-\bar{\alpha}_s \bar{\gamma} \chi'(\bar{\gamma})/N}.$$

In order to obtain the leading behaviour at low x of $F_2(x, Q^2)$ we examine this moment equation near $N = \omega_0$ for which

$$\lim_{N \to \omega_0} R_N = -\frac{\omega_0}{\sqrt{14\bar{\alpha}_s \zeta(3)(N - \omega_0)}}$$

and

$$\lim_{N \to \omega_0} \bar{\gamma} = \frac{1}{2} - \sqrt{\frac{N - \omega_0}{14\bar{\alpha}_s \zeta(3)}}.$$

In this approximation Eq.(6.37) is the Mellin transform of Eq.(6.15). It is important to note that the leading $x^{-\omega_0}$ behaviour arises from the singularity in R_N and is present for *any value of* Q^2. A similar growth at low x arises from the factor $(Q^2/\mu^2)^{\bar{\gamma}}$ but this is only important for $Q^2 \gg \mu^2$.

Furthermore we see from Eq.(6.37) that deviations from Bjorken scaling are present even in the limit of asymptotically large Q^2. The size of these 'anomalous' scaling violations is determined by $\bar{\gamma}$. Accordingly, we call this the **BFKL anomalous dimension** and soon we will show that it is equal to the DGLAP gluon anomalous dimension, γ_{gg}^N (in the low-x, i.e. small-N, limit). From the fact that $\chi(\bar{\gamma}) = N/\bar{\alpha}_s$ and using (for $|\gamma| < 1$)

$$\chi(\gamma) = \frac{1}{\gamma} + 2 \sum_{r=1}^{\infty} \zeta(2r + 1)\gamma^{2r},$$

we can obtain the perturbative expansion of $\bar{\gamma}$:

$$\bar{\gamma} = \frac{\bar{\alpha}_s}{N} + 2\zeta(3) \left(\frac{\bar{\alpha}_s}{N}\right)^4 + 2\zeta(5) \left(\frac{\bar{\alpha}_s}{N}\right)^6 + \mathcal{O}\left(\frac{\bar{\alpha}_s}{N}\right)^7. \qquad (6.38)$$

We are now ready to make the explicit connection with the DGLAP result. At low x, we are not sensitive to the valence quarks so the q and $\bar{q}$ distributions in the proton are equal. Also, in the BFKL treatment we ignored intrinsic quark densities, so we must drop all terms $\propto \Sigma_N(Q^2)$. Finally, the $\ln Q^2$ derivative of the co-efficient functions is sub-leading. Equation (6.26) then becomes

$$\frac{\partial F_{2,N}(Q^2)}{\partial \ln Q^2} = (\langle e_q^2 \rangle 2n_f \gamma_{qg}^N + C_g^N(1, \alpha_s))G_N(Q^2), \qquad (6.39)$$

where

$$\frac{\partial G_N(Q^2)}{\partial \ln Q^2} = \gamma_{gg}^N G_N(Q^2). \qquad (6.40)$$

For fixed coupling this means that

$$G_N(Q^2) = G_N(\mu^2)\left(\frac{Q^2}{\mu^2}\right)^{\gamma_{gg}^N}. \qquad (6.41)$$

Thus, for consistency with Eq.(6.37), $\gamma_{gg}^N = \bar{\gamma}$. Note that the equivalence of the first term in the expansions (in $\bar{\alpha}_s/N$) of γ_{gg}^N and $\bar{\gamma}$ can be seen explicitly by comparing Eq.(6.38) with the $N \to 0$ limit of Eq.(6.28). In addition, we also require that

$$(\langle e_q^2 \rangle 2n_f \gamma_{qg}^N + C_g^N(1, \alpha_s))G_N(\mu^2) = h_{2,N}R_N \mathcal{F}_N^0(\bar{\gamma}, \mu). \qquad (6.42)$$

To summarize, we have shown that the leading twist part of the BFKL solution for the structure function factorizes in a manner consistent with the DGLAP approach. The equivalence of the (leading-twist) BFKL solution and the $N \to 0$ limit of the DGLAP solution allows us to identify the DGLAP gluon anomalous dimension, γ_{gg}^N, with the BFKL anomalous dimension, $\bar{\gamma}$ (calculated to all orders in $\bar{\alpha}_s/N$). The equivalence also allows us to identify the DGLAP coefficient functions with the BFKL 'coefficient function' as in Eq.(6.42). However, there is an ambiguity in extracting the coefficient function which we shall now examine.

In the lowest order of perturbation theory, $\bar{\gamma} = \bar{\alpha}_s/N$, $R_N = 1$, $h_{2,N} = \bar{\alpha}_s \langle e_q^2 \rangle n_f/9$ and $\gamma_{qg}^N = \bar{\alpha}_s/18$. To this order, Eq.(6.39) reduces to

$$\frac{\partial F_{2,N}(Q^2)}{\partial \ln Q^2} = \sum_{q=1}^{n_f} e_q^2 \frac{\bar{\alpha}_s}{9} G_N(Q^2). \qquad (6.43)$$

Comparing with Eq.(6.9), we see that (at this lowest order) the BFKL gluon density defined by Eq.(6.6) is precisely the DGLAP gluon density. Also, comparison with Eq.(6.37) forces us to identify

$$G_N(\mu^2) = \mathcal{F}_N^0(\bar{\gamma}, \mu).$$

However, beyond the lowest order R_N starts to deviate from unity. Since we do not know the scale at which to evaluate α_s we have the freedom to either: absorb all or part of R_N into the input density, $G_N(\mu^2)$, or absorb all or part of R_N into the definition of $\langle e_q^2 \rangle 2n_f \gamma_{qg}^N + C_g^{\prime N}(1, \alpha_s)$. This ambiguity is a factorization scheme ambiguity and is a direct result of the leading logarithmic nature of the calculation.

We know, from the standard renormalization group approach, that it is appropriate to evaluate the anomalous dimensions at the scale Q^2 in the DGLAP evolution. This then forces us to make the replacement,

$$\left(\frac{Q^2}{\mu^2}\right)^{\bar{\gamma}} \to \exp \int_{\mu^2}^{Q^2} \frac{dq^2}{q^2} \bar{\gamma}(q^2) \qquad (6.44)$$

and we have made explicit the fact that $\bar{\gamma}$ should be evaluated at $\alpha_s(q^2)$ on the right hand side. Similarly, we evaluate the coefficient function at $\alpha_s(Q^2)$, i.e. $h_{2,N}(Q^2)$. The scheme ambiguity is still present in R_N, so we have no guidance as to what scale to evaluate it at. The replacement of Eq.(6.44) arises as a result of the radiative corrections which cause the QCD coupling to run. As such it is formally beyond the leading BFKL approximation. For a more detailed investigation of factorization in the high energy regime see the paper by Catani & Hautmann (1994).

6.5 Exclusive distributions and coherence

The derivation of the BFKL equation presented in Chapters 3 and 4 relies upon the validity of the Regge kinematics (i.e. strong ordering in the Sudakov variables). It turns out that this kinematic regime is generally only applicable for the calculation of elastic-scattering cross-sections (and hence total cross-sections), which is where we have been using it hitherto.

In this section we would like to generalize the multi-Regge kinematics so as to allow the calculation of more exclusive quantities, e.g. the number of gluons emitted in deep inelastic scattering. This generalization is made by accounting for QCD coherence effects. Here we present only a brief outline of the motivation for coherence in QCD but refer the reader to the wealth of literature (see e.g. Dokshitzer, Khoze, Troyan & Mueller (1991) and references

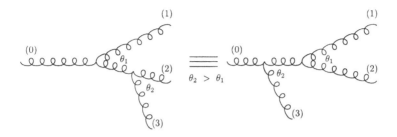

Fig. 6.6. The amplitude for gluon (0) to decay into two gluons (1) and (2) with opening angle θ_1 and for gluon (3) to be radiated off either gluon (1) or gluon (2) (as shown) with opening angle θ_2, where $\theta_2 > \theta_1$, is equivalent to the amplitude for gluon (0) to radiate gluon (3) with opening angle θ_2 and then to decay into gluons (1) and (2).

therein) for a more detailed treatment. We will show that, when accounted for, coherence leads to terms which have additional logarithms (in s) compared to the naive BFKL expectations. These extra logarithms cancel for fully inclusive quantities but not for more exclusive ones (where they provide the dominant contribution). For more details regarding coherence in low-x physics we refer to the work of Ciafaloni (1988), Catani, Fiorani & Marchesini (1990a,b), Catani, Fiorani, Marchesini & Oriani (1991) and Marchesini (1995).

If we consider a time-like (off-shell) parent gluon decaying into two daughter gluons with opening angle θ_1, followed by a further emission of a grand-daughter gluon from one of the daughter gluons with opening angle θ_2, where $\theta_2 > \theta_1$, then at the time of emission the transverse component of the wavelength of the grand-daughter gluon is larger than the transverse spatial separation of the two daughter gluons. In that case the grand-daughter gluon cannot resolve the colours of the individual daughter gluons, but only that of the parent, so that the amplitude for the process is equivalent to the amplitude for the process in which the grand-daughter gluon is emitted directly off the parent, see Fig. 6.6. This is the phenomenon of **colour coherence** and it leads to the angular ordering of sequential gluon emissions in a cascade, i.e. if the opening angle of the ith gluon is θ_i then $\theta_i < \theta_{i-1}$.

In the case of deep inelastic scattering it is convenient to consider successive gluon emission from the target proton, which has zero transverse momentum, up towards the virtual photon, which has momentum **k** transverse to the electron–proton system given by

$$\mathbf{k}^2 = Q^2(1 - y).$$

Suppose the $(i - 1)$th emitted gluon (from the proton) has energy E_{i-1} and that it emits a gluon with a fraction $(1 - z_i)$ of this energy and a transverse momentum of magnitude q_i. The (small) opening angle θ_i of this emitted gluon is given approximately by

$$\theta_i \approx \frac{q_i}{(1 - z_i)E_{i-1}},$$

where z_i is the fraction of the energy of the $(i-1)$th gluon carried off by the ith gluon, i.e.

$$z_i = \frac{E_i}{E_{i-1}}.$$

Colour coherence leads to angular ordering with increasing opening angles towards the hard scale (the photon) so in this case we have $\theta_{i+1} > \theta_i$, which may be expressed as

$$\frac{q_{i+1}}{(1 - z_{i+1})} > \frac{z_i q_i}{(1 - z_i)}. \tag{6.45}$$

In the multi-Regge limit where $z_i, z_{i+1} \ll 1$ this reduces to

$$q_{i+1} > z_i q_i. \tag{6.46}$$

For the first emission, we take $q_0 z_0 \equiv \mu$. The kinematics of the virtual graphs (which reggeize the t-channel gluons) is similarly modified and ensures the cancellation of the collinear singularities in inclusive quantities.

In Chapters 2 and 3 we assumed that the transverse momenta of the gluons in the ladder were all of the same order of magnitude so that the requirement $z_i \ll 1$ meant that the inequality Eq.(6.46) was automatically satisfied. However, we know that we must integrate over *all* transverse momenta of the gluons so that we sample 'corners' of transverse momentum space for which the inequality is violated. As we shall show below, these 'corners' can give rise to super-leading logarithms. These super-leading logarithms cancel when we consider inclusive processes for which we may apply dispersion techniques discussed in Chapter 3, but for certain exclusive processes they do *not* cancel and furnish the leading behaviour as $s \to \infty$.

Before imposing the constraints of angular ordering, it is necessary first to re-write the $t = 0$ BFKL equation in a form which will be suitable for the study of the more exclusive quantities. We can rewrite the $t = 0$ BFKL equation of Eq.(4.13) in the form

$$
\begin{aligned}
f_\omega(\mathbf{k}) = {} & f_\omega^{(0)}(\mathbf{k}) \\
& + \bar{\alpha}_s \int \frac{d^2\mathbf{q}}{\pi \mathbf{q}^2} \int_0^1 \frac{dz}{z} z^\omega \Delta_R(z, k) \Theta(q - \mu) f_\omega(\mathbf{q} + \mathbf{k}),
\end{aligned} \quad (6.47)
$$

where $\mathbf{q} = \mathbf{k}' - \mathbf{k}$ is the transverse momentum of the emitted gluon and the gluon Regge factor is

$$
\ln \Delta_R(z, k) = -\bar{\alpha}_s \int_z^1 \frac{dz'}{z'} \int \frac{d^2\mathbf{q}}{\pi \mathbf{q}^2} \Theta(q - \mu) \Theta(k - q). \quad (6.48)
$$

Equation (6.47) is easy to derive from Eq.(4.13) once we notice that

$$
\ln \Delta_R(z, k) = 2 \ln(1/z) \, \epsilon_G(-k^2) \quad (6.49)
$$

where $1 + \epsilon_G(-k^2)$ is the gluon Regge trajectory derived in Chapter 3. In addition we used the fact that

$$
\frac{1}{\omega - 2\epsilon_G(-k^2)} = \int_0^1 \frac{dz}{z} z^\omega \Delta_R(z, k). \quad (6.50)
$$

The driving term, $f_\omega^{(0)}(\mathbf{k})$, includes the virtual corrections which reggeize the bare gluon. This form of the BFKL equation has a kernel which, under iteration, generates real gluon emissions with all the virtual corrections summed to all orders. As such, it is suitable for the study of the final state. Of course f_ω includes the sum over all final states and as such the μ-dependence cancels between the real and virtual contributions. However, we intend to investigate more exclusive quantities which are no longer infrared finite. The scale μ should then be regarded as the scale above which we can resolve real gluon emission.

Let us now take a specific example. We will look at the contributions to the structure function of an on-shell gluon which come from the emission of either one or two gluons which are constrained to have their transverse momentum less than some scale Q. The energy of the bare on-shell gluon is fixed, thus our boundary condition is

$$F^{(0)}(x, \mathbf{k}) = \delta(x - 1)\,\delta^2(\mathbf{k}), \qquad \text{i.e.}$$
$$f^{(0)}_\omega(\mathbf{k}) = \delta^2(\mathbf{k}). \qquad (6.51)$$

Since the gluon is on shell it does not pick up any corrections due to the reggeization (i.e. we used $\epsilon_G(0) = 0$).

The structure function (defined by integrating over all $q_i^2 \leq Q^2$) thus satisfies the equation,

$$F_\omega(Q^2, \mu^2) = \Theta(Q - \mu) + \sum_{j=1}^{\infty} \prod_{i=1}^{j} \left\{ \bar{\alpha}_s \int \frac{dz_i}{z_i} \frac{d^2\mathbf{q_i}}{\pi \mathbf{q_i}^2} \right.$$
$$\times \quad \left. \Delta_R(z_i, k_i) z_i^\omega \,\Theta(q_i - \mu)\Theta(Q - q_i) \right\}. \quad (6.52)$$

and we have isolated the contributions from i real gluon emissions by iterating the kernel explicitly. Again non-boldface means the modulus of the two-vector.

Ignoring the coherence effects for the moment, the contribution to the structure function from the emission of a single gluon is thus

$$F^{(1)}_\omega(Q^2, \mu^2) = \bar{\alpha}_s \int_{\mu^2}^{Q^2} \frac{d^2\mathbf{q_1}}{\mathbf{q_1}^2} \int_0^1 \frac{dz_1}{z_1} z_1^\omega \Delta_R(z_1, k_1) \qquad (6.53)$$

and $\mathbf{k_1} = -\mathbf{q_1}$ (since the initial gluon is on shell). The Regge factor can then be integrated and yields,

$$\ln \Delta_R(z_1, q_1) = -\bar{\alpha}_s \ln (1/z_1) \ln q_1^2/\mu^2. \qquad (6.54)$$

Let us compute our result as a power series in α_s, i.e. we expand the Regge exponential. Thus

$$F^{(1)}_\omega(Q^2, \mu^2) = \bar{\alpha}_s \int_{\mu^2}^{Q^2} \frac{dq_1^2}{q_1^2} \int_0^1 \frac{dz_1}{z_1} z_1^\omega$$
$$\times \left[1 - \bar{\alpha}_s \ln \frac{1}{z_1} \ln \frac{q_1^2}{\mu^2} + \frac{1}{2}\left(\bar{\alpha}_s \ln \frac{1}{z_1} \ln \frac{q_1^2}{\mu^2} \right)^2 + \mathcal{O}(\alpha_s^3) \right]. (6.55)$$

The z_1-integral can be done by parts and yields

$$F^{(1)}_\omega(Q^2, \mu^2) = \frac{\bar{\alpha}_s}{\omega} \ln \frac{Q^2}{\mu^2} - \frac{1}{2}\left(\frac{\bar{\alpha}_s}{\omega} \ln \frac{Q^2}{\mu^2} \right)^2 + \frac{1}{3}\left(\frac{\bar{\alpha}_s}{\omega} \ln \frac{Q^2}{\mu^2} \right)^3 + \mathcal{O}(\alpha_s^4).$$
$$(6.56)$$

Similarly, the contribution from two-gluon emission is

$$
\begin{aligned}
F_\omega^{(2)}(Q^2,\mu^2) &= \bar{\alpha}_s^2 \int_{\mu^2}^{Q^2} \frac{d^2\mathbf{q_1}}{\pi \mathbf{q_1}^2}\frac{d^2\mathbf{q_2}}{\pi \mathbf{q_2}^2} \int_0^1 \frac{dz_1}{z_1}z_1^\omega \frac{dz_2}{z_2}z_2^\omega [1+\mathcal{O}(\alpha_s)] \\
&= \left(\frac{\bar{\alpha}_s}{\omega}\ln\frac{Q^2}{\mu^2}\right)^2 + \mathcal{O}(\alpha_s^3).
\end{aligned} \tag{6.57}
$$

In fact, a more detailed treatment (Marchesini (1995)) reveals that the inclusive structure function satisfies

$$
F_\omega(Q^2) \equiv \sum_{i=0}^\infty F_\omega^{(i)}(Q) = \left(\frac{Q^2}{\mu^2}\right)^{\bar{\gamma}}, \tag{6.58}
$$

where $\bar{\gamma}$ is the BFKL anomalous dimension.

As we alluded to at the start of this section – these results (with the exception of Eq.(6.58)) are wrong. We must modify Eqs.(6.47) and (6.48) to account for coherence, so that Eq.(6.52) becomes

$$
\begin{aligned}
F_\omega(Q^2,\mu^2) &= \Theta(Q-\mu) + \sum_{j=1}^\infty \prod_{i=1}^j \left\{ \bar{\alpha}_s \int \frac{dz_i}{z_i}\frac{d^2\mathbf{q_i}}{\pi \mathbf{q_i}^2} \right. \\
&\quad \left. \times \Delta(z_i,k_i)z_i^\omega\, \Theta(q_{i+1}-z_iq_i)\Theta(Q-q_i) \right\},
\end{aligned} \tag{6.59}
$$

where the coherence improved Regge factor is

$$
\ln\Delta(z_i,k_i,q_i) = -\int_{z_i}^1 \frac{dz}{z}\int \frac{d^2\mathbf{q}}{\pi\mathbf{q}^2}\bar{\alpha}_s\Theta(q-z_iq_i)\Theta(k_i-q). \tag{6.60}
$$

Let us now re-compute the single gluon emission cross-section. The Regge factor now becomes (because $k_1 = q_1$)

$$
\ln\Delta(z_1,k_1,q_1) = -\bar{\alpha}_s\ln^2(1/z). \tag{6.61}
$$

Expanding as a power series in α_s we now obtain

$$
F_\omega^{(1)}(Q^2,\mu^2) = \bar{\alpha}_s \int_{\mu^2}^{Q^2} \frac{dq_1^2}{q_1^2}\int_0^{Q/k}\frac{dz_1}{z_1}z_1^\omega\left[1 \quad \bar{\alpha}_s\ln^2\frac{1}{z}+\mathcal{O}(\alpha_s^2)\right]. \tag{6.62}
$$

The z-integrals can again be performed using

$$
\begin{aligned}
\int_0^{Q/k}\frac{dz}{z}z^\omega &= \frac{1}{\omega}+\mathcal{O}(1), \\
\int_0^{Q/k}\frac{dz}{z}z^\omega\ln^2 z &= \frac{2}{\omega^3}+\mathcal{O}(1),
\end{aligned} \tag{6.63}
$$

and we neglect those terms which are not singular in the limit $\omega \to 0$ (which corresponds to keeping those terms which are leading in the Regge limit and beyond). Our final answer for the coherence improved calculation of the single gluon emission rate is therefore

$$F_\omega^{(1)}(Q^2, \mu^2) = \left[\frac{\bar{\alpha}_s}{\omega} \ln \frac{Q^2}{\mu^2} - 2 \frac{\bar{\alpha}_s^2}{\omega^3} \ln \frac{Q^2}{\mu^2} + \cdots \right]. \qquad (6.64)$$

Inverting the Mellin transform, we therefore see that the cross-section for single gluon emission is enhanced over the naive BFKL expectation by a factor of $\ln s$. This logarithm (and those that occur at higher orders in α_s) must be cancelled in the inclusive sum. We can therefore write the cross-section for two-gluon emission, i.e.

$$F_\omega^{(2)}(Q^2, \mu^2) = \left[\frac{1}{2} \left(\frac{\bar{\alpha}_s}{\omega} \ln \frac{Q^2}{\mu^2} \right)^2 + 2 \frac{\bar{\alpha}_s^2}{\omega^3} \ln \frac{Q^2}{\mu^2} + \cdots \right]. \qquad (6.65)$$

Although we expect coherence to affect the details of the final state dramatically, it also generates sub-leading corrections to the inclusive BFKL cross-section. These corrections are embodied in the solution to Eq.(6.59) and have been studied in the work of Kwiecinski, Martin & Sutton (1995).

Before finishing this chapter, a few words are in order regarding other processes that allow a study of the BFKL (hard) Pomeron in the $t = 0$ limit. In Section 6.3 we considered the associated jet production in deep inelastic scattering. By now, it should be clear that a similar process can be studied in hadron collider experiments (or in γ–p collisions with nearly on-shell photons), namely, events containing two jets which are produced so that they are separated by a large interval in rapidity (i.e. **double associated jet production**) (Mueller & Navalet (1987), Del Duca & Schmidt (1994a,b), Stirling (1994)). The hadron impact factors of Eq.(6.16) and Eq.(6.17) are then applied to each hadron-jet vertex. Similarly, rather than insisting on the production of a forward jet (or forward and backward jet pair) we could look for heavy quark production in these rapidity regions. The quark mass then provides the large scale in the impact factor(s) (see e.g. Catani, Ciafaloni & Hautmann (1990, 1991), Collins & Ellis (1991) and Levin, Ryskin, Shabelskii & Shuvaev (1991)).

6.6 Summary

• The high energy limit of deep inelastic scattering corresponds to the limit of low Bjorken-x. The leading log $1/x$ approximation leads to the low-x behaviour which is characterized by the leading eigenvalue of the BFKL kernel, i.e. $F_2(x, Q^2) \sim x^{-\omega_0}(Q^2)^{1/2}$.

• The diffusion properties of the BFKL equation mean that a large contribution to the total deep inelastic cross-section can arise from the non-perturbative domain where the typical transverse momenta are not large.

• A process which is better suited to the application of perturbative QCD is that of associated jet production. The observation of an additional jet, travelling close to the direction defined by the incoming hadron, ensures the clean factorization of the non-perturbative dynamics into known parton distribution functions.

• The more conventional DGLAP formalism of deep inelastic scattering can be related to the BFKL approach. The leading twist contribution can be extracted from the BFKL calculation and can be shown to be equivalent to the soft gluon limit of the DGLAP equations (i.e. the limit in which only the singular parts of the all-orders DGLAP splitting functions are kept).

• The multi-Regge kinematics (i.e. the strong ordering of the longitudinal momentum fractions) used to compute the elastic-scattering amplitudes (and hence total inclusive cross-sections) via the BFKL equation is inappropriate for the consideration of more exclusive quantities. Coherence effects lead to additional logarithms in energy which only cancel in the inclusive sum.

6.7 Appendix A

In this appendix we compute the virtual photon impact factor required to compute the deep inelastic structure functions. In Fig. 6.7 we show two of the four diagrams which are needed (the other two are trivially obtained by reversing the direction of the quark line). Our calculation is very much analogous to that of Section 4.4.

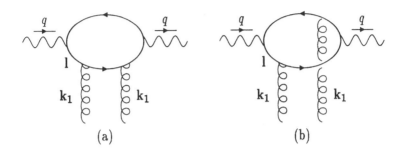

Fig. 6.7. Two of the four graphs used to compute the photon impact factor.

For Fig. 6.7(b), we have the amplitude

$$A_{(b)}^{\mu\nu\alpha\beta} = -(4\pi)^2 \alpha_s \alpha\, e_q^2 \frac{\mathrm{Tr}(\hat{l}\gamma^\mu(\hat{l} - \hat{k}_1)\gamma^\beta(\hat{l} - \hat{k}_1 - \hat{q})\gamma^\nu(\hat{l} - \hat{q})\gamma^\alpha)}{l^2(l - k_1 - q)^2}.$$

$$(\mathrm{A.6.1})$$

We use the notation $\hat{l}$ for $\gamma_\mu l^\mu$ and e_q is the quark charge in units of the proton charge. As in Section 4.4 we have factored out the colour factor. We express the four momenta l^μ, k_1^μ and q^μ in terms of the light-like vectors p_1^μ and p_2^μ ($p_2^\mu \equiv p^\mu$ is the incoming proton momentum) and their transverse components, i.e.

$$
\begin{aligned}
l^\mu &= \rho p_1^\mu + \lambda p_2^\mu + l_\perp^\mu \\
k_1^\mu &= \rho_1 p_1^\mu + \lambda_1 p_2^\mu + k_{1\perp}^\mu \\
q^\mu &= p_1^\mu - x p_2^\mu.
\end{aligned}
$$

The denominator factors are also as in Section 4.4, i.e.

$$\frac{1}{l^2}\frac{1}{(l - k_1 - q)^2} = \frac{\rho(1 - \rho)}{D_1 D_2},$$

$$(\mathrm{A.6.2})$$

where

$$
\begin{aligned}
D_1 &= \mathbf{l}^2 + Q^2 \rho(1 - \rho) \\
D_2 &= (\mathbf{l} - \mathbf{k_1})^2 + Q^2 \rho(1 - \rho)
\end{aligned}
$$

(recall $Q^2 = -q^2 > 0$).

In the high energy limit we are only interested in those terms which are proportional to $p_1^\mu p_1^\nu$, i.e.

$$A_{(b)}^{\mu\nu\alpha\beta} = -(4\pi)^2 \alpha \alpha_s e_q^2 \frac{\rho(1 - \rho)}{D_1 D_2}(A_{(b)}^{\alpha\beta} p_1^\mu p_1^\nu + ...)$$

$$(\mathrm{A.6.3})$$

and the tensor associated with the $p_1^\mu p_1^\nu$ factor is given by

$$A_{(b)}^{\alpha\beta} = \frac{4}{W^4} \text{Tr}[\hat{l}\hat{p}_2(\hat{l} - \hat{k}_1)\gamma^\beta(\hat{l} - \hat{k}_1 - \hat{q})\hat{p}_2(\hat{l} - \hat{q})\gamma^\alpha]. \quad \text{(A.6.4)}$$

We have used the fact that $2q \cdot p \approx W^2$ in the high energy limit.

Also, we need to contract with the photon polarization vectors. For longitudinal photons we contract with

$$\epsilon_\alpha^L \epsilon_\beta^{L*} = \frac{4Q^2}{W^2} \frac{p_{2\alpha} p_{2\beta}}{W^2} + \cdots, \quad \text{(A.6.5)}$$

whilst for transverse photons we need to contract with

$$\sum \epsilon_\alpha^T \epsilon_\beta^{T*} = -g_{\alpha\beta} + \frac{4Q^2}{W^2} \frac{p_{2\alpha} p_{2\beta}}{W^2} + \cdots. \quad \text{(A.6.6)}$$

The dots refer to terms which ultimately vanish since they contain factors $\sim q_\alpha q_\beta$ and $\sim q_\alpha p_{2\beta} + q_\beta p_{2\alpha}$. Such terms vanish since current conservation implies that their contraction with the leptonic tensor must vanish. Thus, on contracting the relevant components of the trace it follows (without too much work) that

$$-g_{\alpha\beta} A_{(b)}^{\alpha\beta} = 16[\mathbf{l} \cdot (\mathbf{l} - \mathbf{k_1}) + \mathbf{k_1^2} \rho(1 - \rho)] \quad \text{(A.6.7)}$$
$$= 8[D_1 + D_2 - 2Q^2 \rho(1 - \rho) - \mathbf{k_1^2}(\rho^2 + (1 - \rho)^2)]$$

and

$$\frac{4Q^2}{W^2} \frac{p_{2\alpha} p_{2\beta}}{W^2} A_{(b)}^{\alpha\beta} = 32Q^2 \rho^2 (1 - \rho)^2. \quad \text{(A.6.8)}$$

Thus, in the same notation as in Eqs.(4.38) and (4.39), we have the following contribution to the amplitudes from Fig. 6.7(b):

$$A_{(b)\Sigma}^{\mu\nu} = -(4\pi)^2 \alpha_s \alpha \, e_q^2 \frac{8\rho(1 - \rho)}{D_1 D_2} p_1^\mu p_1^\nu$$
$$\times \left\{ D_1 + D_2 - 2Q^2 \rho(1 - \rho) - \mathbf{k_1^2}[\rho^2 + (1 - \rho)^2] \right\} \quad \text{(A.6.9)}$$

and

$$A_{(b)L}^{\mu\nu} = -(4\pi)^2 \alpha_s \alpha \, e_q^2 \frac{32\rho(1 - \rho)}{D_1 D_2} p_1^\mu p_1^\nu \left\{ Q^2 \rho^2 (1 - \rho)^2 \right\}. \quad \text{(A.6.10)}$$

and $A_{(b)\Sigma}^{\mu\nu}$ is defined such that the amplitude for scattering transverse photons is

$$(A_{(b)\Sigma}^{\mu\nu} + A_{(b)L}^{\mu\nu})/2.$$

We can now follow the steps of Section 4.4 to deduce the corresponding contributions to the impact factors:

$$\Phi^{\Sigma}_{(b)}(\mathbf{k_1}) = -8\alpha\alpha_s \sum_{q=1}^{n_f} e_q^2 \int d\rho \, d^2\mathbf{l}$$

$$\times \left\{ \frac{1}{D_1} + \frac{1}{D_2} - \frac{2Q^2\rho(1-\rho) + \mathbf{k_1^2}[\rho^2 + (1-\rho)^2]}{D_1 D_2} \right\} \text{(A.6.11)}$$

and

$$\Phi^{L}_{(b)}(\mathbf{k_1}) = -32\alpha\alpha_s \sum_{q=1}^{n_f} e_q^2 \int d\rho \, d^2\mathbf{l} \left\{ Q^2 \frac{\rho^2(1-\rho)^2}{D_1 D_2} \right\}. \quad \text{(A.6.12)}$$

A factor of 2 has been included to account for the related graph which has the quark line circulating in the opposite direction, and we have summed over all n_f flavours of quark.

Fortunately, we do not need to do any more work in order to extract the contribution from the graph shown in Fig. 6.7(a). It is related to the above impact factor via

$$\Phi^{\lambda}_{(a)}(\mathbf{k}) = -\Phi^{\lambda}_{(b)}(0). \quad \text{(A.6.13)}$$

Using this, and noting that

$$\int d^2\mathbf{l} \frac{1}{D_1^n} = \int d^2\mathbf{l} \frac{1}{D_2^n}, \quad \text{(A.6.14)}$$

we can write the complete impact factors for deep inelastic scattering as

$$\Phi_{\Sigma}(\mathbf{k_1}) = 8\alpha\alpha_s \sum_{q=1}^{n_f} e_q^2 \int d\rho \, d^2\mathbf{l}$$

$$\times \left\{ \frac{\mathbf{k_1^2}[\rho^2 + (1-\rho)^2]}{D_1 D_2} - Q^2\rho(1-\rho) \left(\frac{1}{D_1} - \frac{1}{D_2} \right)^2 \right\} \text{(A.6.15)}$$

and

$$\Phi_L(\mathbf{k_1}) = 16\alpha\alpha_s \sum_{q=1}^{n_f} e_q^2 \int d\rho \, d^2\mathbf{l} \left\{ Q^2\rho^2(1-\rho)^2 \left(\frac{1}{D_1} - \frac{1}{D_2} \right)^2 \right\}.$$

$$\text{(A.6.16)}$$

The transverse polarization impact factor, Φ_T, is given by

$$\Phi_T = \frac{1}{2}(\Phi_{\Sigma} + \Phi_L). \quad \text{(A.6.17)}$$

Using Eq.(6.2), we can now deduce the impact factor for the sum of transverse and longitudinal cross-sections:

$$\Phi_2(\mathbf{k_1}) = \Phi_T(\mathbf{k_1}) + \Phi_L(\mathbf{k_1}) = \frac{\Phi_\Sigma(\mathbf{k_1}) + 3\Phi_L(\mathbf{k_1})}{2}. \quad \text{(A.6.18)}$$

As in Section 4.4, we can go further and perform the l integral at the expense of introducing a Feynman parameter, τ. We need to use

$$\int \frac{d^2 \mathbf{l}}{D_1^2} = \int \frac{d^2 \mathbf{l}}{D_2^2} = \frac{\pi}{\rho(1-\rho)Q^2} \quad \text{(A.6.19)}$$

and

$$\int \frac{d^2 \mathbf{l}}{D_1 D_2} = \int_0^1 d\tau \frac{\pi}{\rho(1-\rho)Q^2 + \tau(1-\tau)\mathbf{k_1^2}}. \quad \text{(A.6.20)}$$

After some simple algebra, we then obtain

$$\Phi_T(\mathbf{k_1}) = 4\pi\alpha\alpha_s \sum_{q=1}^{n_f} e_q^2 \int_0^1 d\rho d\tau \frac{\mathbf{k_1^2}}{\rho(1-\rho)Q^2 + \tau(1-\tau)\mathbf{k_1^2}}$$
$$\times [\tau^2 + (1-\tau)^2][\rho^2 + (1-\rho)^2] \quad \text{(A.6.21)}$$

and

$$\Phi_L(\mathbf{k_1}) = 32\pi\alpha\alpha_s \sum_{q=1}^{n_f} e_q^2 \int_0^1 d\rho d\tau \frac{\mathbf{k_1^2}}{\rho(1-\rho)Q^2 + \tau(1-\tau)\mathbf{k_1^2}}$$
$$\times [\rho(1-\rho)\tau(1-\tau)]. \quad \text{(A.6.22)}$$

6.8 Appendix B

The saddle point method of integration is a powerful tool for approximating integrals which may be cast into the form

$$\int_{-\infty}^{\infty} dx\, g(x)e^{-f(x)}. \quad \text{(B.6.1)}$$

The method is valid provided the function $f(x)$ has a minimum at some value $x = x_0$ and that it is 'very convex' in that region. This means that the nth derivative, $f^{(n)}(x)$, of $f(x)$ obeys the inequality

$$f^{(n)}(x_0) \ll \left(f^{(2)}(x_0)\right)^{n/2},$$

so that $f(x)$ may be approximated by

$$f(x) \approx f(x_0) + \frac{1}{2}f^{(2)}(x_0)(x - x_0)^2 \quad \text{(B.6.2)}$$

(the first derivative vanishes at $x = x_0$ since $f(x)$ has a minimum there).

Furthermore, the function $g(x)$ is assumed to be a 'slowly varying' function at $x = x_0$. This means that it may be approximated by its value at $x = x_0$ or, in cases where that value vanishes, by its first non-vanishing even order derivative at $x = x_0$, i.e. if the first non-vanishing even order derivative at $x = x_0$ is the $(2m)$th derivative then we write

$$g(x) \approx \frac{1}{(2m)!} g^{(2m)}(x_0) (x - x_0)^{2m} . \qquad (B.6.3)$$

Substituting Eqs.(B.6.2) and (B.6.3) into Eq.(B.6.1) and changing variables to $y = (x - x_0)$ we obtain the Gaussian integral

$$e^{-f(x_0)} \int_{-\infty}^{\infty} dy \, \frac{1}{(2m)!} g^{(2m)}(x_0) y^{2m} \exp\left(-\frac{1}{2} f^{(2)}(x_0) y^2\right)$$
$$= \frac{\sqrt{2\pi}}{2^m m!} \frac{g^{(2m)}(x_0)}{(f^{(2)}(x_0))^{m+1/2}} e^{-f(x_0)} . \qquad (B.6.4)$$

The corrections to the above approximation are of order

$$\frac{f^{(n)}(x_0)}{(f^{(2)}(x_0))^{n/2}} \quad \text{or} \quad \frac{g^{(2m+2)}(x_0)}{g^{(2m)}(x_0) f^{(2)}(x_0)} .$$

7

Diffraction

In the preceding chapter, we focused on some interesting total cross-sections. That is, we were concerned with the behaviour of the (imaginary part of the) scattering amplitudes in the forward direction (i.e. $t = 0$). It is now time to turn our attention to processes which involve the square of the scattering amplitude. Since in the Regge limit the centre-of-mass energy is much larger than the momentum transferred from the incoming particles to any of the outgoing particles such processes must produce a rapidity gap (see Section 1.10) in the final state.

After a brief word regarding elastic scattering at $t = 0$ we continue by looking at processes at large t. Of course we will find a high energy behaviour which is driven by the leading eigenvalue of the BFKL kernel. In addition, we demonstrate that large t is a good way of keeping the dynamics perturbative (recall that the impact factors were the only way to ensure this in the $t = 0$ case) and that the dominant contributions are characterized by the physics of diffusion in the transverse plane. After demonstrating these important points, we go on to discuss the specific example of vector meson production in two-photon collisions, i.e. $\gamma\gamma \to VV$ where V denotes a vector meson.

The second part of this chapter will be concerned with the physics of diffraction dissociation. In particular, we look in some detail at the particular process of photon dissociation in deep inelastic scattering. By working in the proton (target) rest frame we will be able to discuss the process in a way which is appealing to our physical intuition.

7.1 Elastic scattering at $t = 0$

At $t = 0$ we looked, in the preceding chapter, at the specific example of the forward Compton amplitude, $\gamma p \to \gamma p$. Of course this

175

amplitude also gives us the corresponding differential distribution, $d\sigma/dt$, for the elastic processes at $t = 0$ via

$$\left.\frac{d\sigma}{dt}\right|_{t=0} = \frac{|A(s,0)|^2}{16\pi s^2}. \tag{7.1}$$

A similar process, with a higher rate (than the Compton process), is that of $\gamma p \to V p$ where V is a vector meson and the photon can be real or virtual. For real photons, there is the possibility of using perturbation theory provided the meson is heavy enough. For virtual photons one can study the production of both light and heavy mesons. As well as acting as a possible probe of the perturbative dynamics, these processes allow important information to be extracted about the physics that determines the γV impact factor, which cannot be computed purely within perturbation theory. There has been much interest in this process and here we merely refer to the original papers by Ryskin (1993), Brodsky *et al.* (1994) and the review by Abramowicz, Frankfurt & Strikman (1995).

7.2 Diffusion in large t elastic scattering

In Chapter 4, we derived an expression for the elastic-scattering amplitude at large t (see Eq.(4.52)). We could now proceed to convolute the universal four-point function of Eq.(4.52) with some appropriate impact factors in order to compute the physical cross-sections. However, we need first to establish the circumstances under which perturbation theory ought to apply. Recall the discussion of Section 5.1, where (for $t = 0$) it was demonstrated that the typical transverse momenta at some point inside the Pomeron are governed by the scales within the impact factors, with a distribution characterized by the diffusion equation, Eq.(5.1). We would now like to make a similar study for the case of non-zero t.

In terms of the energy variable, $y = \ln s/k^2$, the generic scattering amplitude can be written (see Eq.(4.36))

$$\frac{\Im m\, A(s,t)}{s} = \frac{\mathcal{G}}{(2\pi)^4} \int d^2\mathbf{k_1} d^2\mathbf{k_2}\, \frac{F(y,\mathbf{k_1},\mathbf{k_2},\mathbf{q})}{\mathbf{k_2^2}(\mathbf{k_1} - \mathbf{q})^2}$$
$$\times\, \Phi_A(\mathbf{k_1},\mathbf{q})\Phi_B(\mathbf{k_2},\mathbf{q}), \tag{7.2}$$

where Φ_A and Φ_B are the impact factors for the Pomeron coupling to the external particles and the universal four-point function is given by

$$
\frac{F(y, \mathbf{k_1}, \mathbf{k_2}, \mathbf{q})}{k_2^2 (\mathbf{k_1} - \mathbf{q})^2} = \frac{1}{(2\pi)^6} \int d^2 \mathbf{b_1} d^2 \mathbf{b_1'} d^2 \mathbf{b_2} d^2 \mathbf{b_2'}
$$
$$
\times \, e^{-i[\mathbf{k_1} \cdot \mathbf{b_1} + (\mathbf{q} - \mathbf{k_1}) \cdot \mathbf{b_1'} - \mathbf{k_2} \cdot \mathbf{b_2} - (\mathbf{q} - \mathbf{k_2}) \cdot \mathbf{b_2'}]}
$$
$$
\times \int_{-\infty}^{\infty} d\nu \frac{\nu^2}{(\nu^2 + 1/4)^2} e^{\bar{\alpha}_s \chi_0(\nu) y} \tilde{\phi}_0^\nu (\mathbf{b_1}, \mathbf{b_1'}, 0) \tilde{\phi}_0^{\nu *} (\mathbf{b_2}, \mathbf{b_2'}, 0). \tag{7.3}
$$

This equation has been obtained from Eqs.(4.46) and (4.52) after a change of variables to eliminate the c-dependence and after taking the (leading) $n = 0$ approximation.

To investigate the internal dynamics of the Pomeron, it is convenient to introduce the functions,

$$
\psi_A(y, \mathbf{r}, \mathbf{q}) = \int \frac{d^2 \mathbf{k}}{(2\pi)^2} e^{-i\mathbf{k} \cdot \mathbf{r}} d^2 \mathbf{k_1} \frac{F(y, \mathbf{k_1}, \mathbf{k}, \mathbf{q})}{\mathbf{k}^2 (\mathbf{k_1} - \mathbf{q})^2} \Phi_A(\mathbf{k_1}, \mathbf{q}) \tag{7.4}
$$

and

$$
\psi_B^*(0, \mathbf{r}, \mathbf{q}) = \int d^2 \mathbf{k_2} e^{i\mathbf{k_2} \cdot \mathbf{r}} \Phi_B(\mathbf{k_2}, \mathbf{q}). \tag{7.5}
$$

These two functions can be thought of as impact factors in impact parameter space ($\mathbf{r}$ is the impact parameter conjugate to the internal momentum, $\mathbf{k}$ and can be thought of as the 'transverse size' of the Pomeron), i.e.

$$
\frac{\Im m \, A(s, t)}{s} = \frac{\mathcal{G}}{(2\pi)^4} \int d^2 \mathbf{r} \, \psi_A(y, \mathbf{r}, \mathbf{q}) \psi_B^*(0, \mathbf{r}, \mathbf{q}). \tag{7.6}
$$

Note that all the BFKL dynamics is subsumed into ψ_A but that, as in Eq.(5.2), we are free to partition the energy dependence as we choose. Also note that these 'impact factors' have different dimensions (ψ_A has dimensions of an area whilst ψ_B has dimensions of an inverse area). Equation (7.6) is shown graphically in Fig. 7.1.

We now wish to focus on the r-dependence of ψ_A as y varies. Physically, we are looking to see what are the typical separations of the two gluons which couple into the lower impact factor (since $\mathbf{r} = \mathbf{b_2'} - \mathbf{b_2}$). We shall show that the largeness of the momentum transfer, $-t = \mathbf{q}^2$, is sufficient to keep this distance small (and hence support the use of perturbation theory) regardless of the size of the external particles.

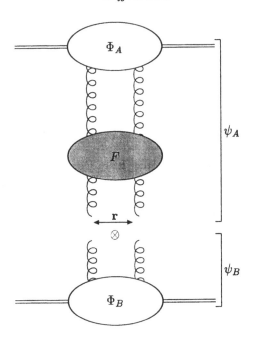

Fig. 7.1. Graphical representation of Eq.(7.6). The convolution represents the integral over the Pomeron transverse size, **r**.

Using Eq.(4.49) for $\tilde{\phi}_0^\nu$ we can combine Eqs.(7.3) and (7.4) to write

$$\psi_A(y,\mathbf{r},\mathbf{q}) = \frac{1}{(2\pi)^6} \int d^2\mathbf{R} \int d\nu \frac{\nu^2}{(\nu^2 + 1/4)^2} e^{\bar{\alpha}_s \chi_0(\nu) y}$$

$$\times V_\nu(\mathbf{q}, Q) \left[\frac{\mathbf{r}^2}{(\mathbf{R} - \mathbf{r}/2)^2 (\mathbf{R} + \mathbf{r}/2)^2} \right]^{1/2 - i\nu} e^{i\mathbf{q}\cdot(\mathbf{R} - \mathbf{r}/2)},$$

$$(7.7)$$

where the impact factor dependent term is

$$V_\nu(\mathbf{q}, Q) = \int d^2\mathbf{k}_1 d^2\mathbf{b}_1 d^2\mathbf{b}_1' e^{-i[\mathbf{k}_1 \cdot \mathbf{b}_1 + (\mathbf{q} - \mathbf{k}_1)\cdot \mathbf{b}_1']}$$

$$\times \Phi_A(\mathbf{k}_1, \mathbf{q}) \left[\frac{(\mathbf{b}_1 - \mathbf{b}_1')^2}{\mathbf{b}_1{}^2 \mathbf{b}_1'{}^2} \right]^{1/2 + i\nu}. \qquad (7.8)$$

Note that we have changed variables from $\mathbf{b}_2$ and $\mathbf{b}_2'$ to

$\mathbf{R} = (\mathbf{b_2} + \mathbf{b_2'})/2$ and the $\mathbf{k}$-integral in Eq.(7.4) gives a delta function which fixes $\mathbf{r} = \mathbf{b_2'} - \mathbf{b_2}$. Also, we have written explicitly the dependence of V_ν upon the scale Q, which characterizes the size of the particle A.

The fact that the eigenfunctions of the kernel are no longer simple powers of the momentum mean that we must face up to the rather unwieldy nature of these expressions. However, it is possible to perform the two-dimensional $\mathbf{R}$-integral by introducing a Feynman parameter, x, and using standard integrals (see e.g. Gradshteyn & Ryzhik (1994)). This gives

$$\psi_A(y, \mathbf{r}, \mathbf{q}) = \frac{r}{(2\pi)^5} \int_{-\infty}^{\infty} d\nu \frac{\nu^2}{(\nu^2 + 1/4)^2} e^{\bar{\alpha}_s \chi_0(\nu) y} V_\nu(\mathbf{q}, Q)$$

$$\times \frac{1}{\Gamma^2(1/2 - i\nu)} \int_0^1 \frac{dx \, e^{-i\mathbf{q} \cdot \mathbf{r} \, x}}{\sqrt{x(1-x)}} \left(\frac{\mathbf{q}^2}{4}\right)^{-i\nu} K_{2i\nu}(qr\sqrt{x(1-x)}), (7.9)$$

where $K_{2i\nu}(qr\sqrt{x(1-x)})$ is a modified Bessel function (see, e.g. Abramowitz & Stegun (1972)) and, as usual, non-boldface is used to denote the modulus of the two-vectors.

Subsequent development clearly necessitates that we say something about the impact factor. However, the presence of the $(\mathbf{q}^2)^{-i\nu}$ factor allows us to recognize that a similar factor must be present in V_ν. In particular, we consider the simplest case where the impact factor is pointlike (i.e. has no scale, Q). Thus we take

$$V_\nu(\mathbf{q}, Q) = (\mathbf{q}^2/4)^{-1/2 + i\nu}.$$

We can now write

$$\psi_A(y, \mathbf{r}, \mathbf{q}) = \frac{1}{(2\pi)^5} \int d\nu \frac{\nu^2}{(\nu^2 + 1/4)^2} e^{\bar{\alpha}_s \chi_0(\nu) y} \frac{2r/q}{\Gamma^2(1/2 - i\nu)}$$

$$\times \int_0^1 \frac{dx}{\sqrt{x(1-x)}} e^{-i\mathbf{q} \cdot \mathbf{r} x} K_{2i\nu}(qr\sqrt{x(1-x)}). \quad (7.10)$$

In general, the $\mathbf{r}$ angular integral is non-trivial when performing the convolution with the ψ_B impact factor. However, we are presently interested in the typical values of the modulus of $\mathbf{r}$ within ψ_A. As such we consider the angular integrated quantity

$$\bar{\psi}_A(y,r,q) \quad = \quad \frac{1}{(2\pi)^4}\int d\nu \frac{\nu^2}{(\nu^2+1/4)^2}e^{\bar{\alpha}_s\chi_0(\nu)y}\frac{2r/q}{\Gamma^2(1/2-i\nu)}$$

$$\times \int \frac{dx}{\sqrt{x(1-x)}}J_0(qrx)K_{2i\nu}(qr\sqrt{x(1-x)}), \quad (7.11)$$

where $J_0(qrx)$ is a Bessel function.

For $qr \lesssim 1$ we can use the small argument expansion of the Bessel functions. In which case,

$$\bar{\psi}_A(y,r,q) \quad \approx \quad \frac{1}{(2\pi)^4}\int d\nu \frac{\nu^2}{(\nu^2+1/4)^2}e^{\bar{\alpha}_s\chi_0(\nu)y}\int \frac{dx}{\sqrt{x(1-x)}}$$

$$\times \left[\frac{[qr/2\sqrt{x(1-x)}]^{2i\nu}}{\Gamma(1+2i\nu)} - \frac{[qr/2\sqrt{x(1-x)}]^{-2i\nu}}{\Gamma(1-2i\nu)}\right]$$

$$\times \frac{2r/q}{\Gamma^2(1/2-i\nu)}\frac{i\pi}{2\sinh 2\pi\nu}. \quad (7.12)$$

The x-integral can now be performed and after taking the limit of small ν (which, as usual, will give the dominant contribution for large enough y) it can be shown that

$$\bar{\psi}_A(y,r,q) \approx \frac{1}{\pi^4}\int d\nu\, \nu e^{\omega_0 y-a^2 y\nu^2}\sin\left(\nu\ln\frac{16}{q^2r^2}\right)\frac{r}{q}, \quad (7.13)$$

where $a^2 = 14\bar{\alpha}_s\zeta(3)$. Performing the ν-integral then yields our final result, i.e.

$$\bar{\psi}_A(y,r,q) \approx \frac{1}{2\pi^4}\frac{\sqrt{\pi}r}{q}\frac{e^{\omega_0 y}}{(a^2 y)^{3/2}}\xi\exp\left(-\frac{\xi^2}{4a^2 y}\right), \quad (7.14)$$

where $\xi \equiv \ln(16/q^2r^2)$. Notice that $\bar{\psi}_A/r$ is also a solution to the diffusion equation of Section 5.1, i.e. Eq.(5.1).

For $qr \gg 1$, the x-integral is dominated by the end-point regions (close to 0 and 1). We can then approximate Eq.(7.11) by

$$\bar{\psi}_A(y,r,q) \quad \approx \quad \frac{1}{(2\pi)^4}\int d\nu \frac{\nu^2}{(\nu^2+1/4)^2}e^{\bar{\alpha}_s\chi_0(\nu)y}\frac{2r/q}{\Gamma^2(1/2-i\nu)}$$

$$\times \int_0^\infty \frac{dx}{\sqrt{x}}K_{2i\nu}(qr\sqrt{x})[1+J_0(qr)]. \quad (7.15)$$

Note that we can concentrate on the $x \to 0$ end-point (since the $x \to 1$ contribution is identical) and the upper limit on the

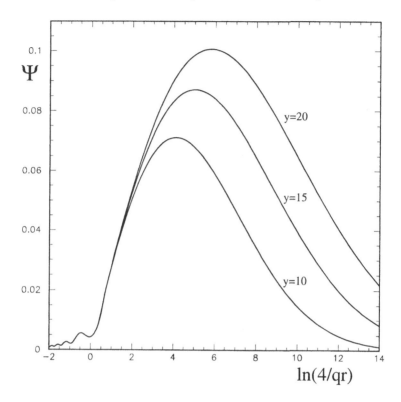

Fig. 7.2. The function Ψ as a function of $\ln(4/qr)$ $(= \xi/2)$ at different y values.

x-integral can be approximated by infinity within our approximations. Equation (7.15) can be integrated about the saddle point at $\nu = 0$ to yield

$$\bar{\psi}_A(y,r,q) \approx \frac{1}{(2\pi)^4} \frac{16\sqrt{\pi}r}{q} \frac{e^{\omega_0 y}}{(a^2 y)^{3/2}} \frac{1}{qr}[1 + J_0(qr)]. \quad (7.16)$$

Note that there is no diffusion into (or from) this region.

In Fig. 7.2 we plot the ξ-dependence of

$$\Psi \equiv \bar{\psi}_A \frac{\pi q}{2r} \left[\frac{e^{\omega_0 y}}{(a^2 y)^{3/2}}\right]^{-1},$$

i.e. we have divided out the typical energy dependent factors to allow a clear demonstration of the diffusion properties. It clearly illustrates the dominance of the region $\xi \gtrsim 0$. Notice that for $\xi > 0$

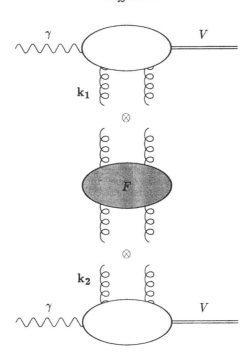

Fig. 7.3. The quasi-elastic scattering process $\gamma\gamma \to VV$.

there is diffusion, i.e. the width in ξ increases with increasing y, indicating that as the energy increases a larger range of the Pomeron transverse size is important. On the other hand for $\xi < 0$ there is no change in shape as y varies. Indeed, the contribution for $\xi < 0$ is very small. We see that the momentum transfer acts as a dividing scale between the region of diffusion (which is dominant) and the scaling region (where the contribution is small). So, for large enough $-t$ we can be sure that the dominant contribution arises from small values of the Pomeron transverse size for which the QCD coupling is small. In this region we expect perturbation theory to be valid.

To make these features more explicit, let us look at a specific example. Namely, we consider the process shown in Fig. 7.3, i.e. $\gamma\gamma - VV$, where V is a vector meson. The incoming photons are assumed to be on shell. A suitable model for the impact factor is to take

$$\Phi_A(\mathbf{k_1}, \mathbf{q}) = C \left(\frac{1}{m^2 + (\mathbf{k} - \mathbf{q}/2)^2} - \frac{1}{m^2 + \mathbf{q}^2/4} \right), \qquad (7.17)$$

where C is a dimensionful constant and depends upon the mass (m) and decay width of the vector meson state. This is something of a toy impact factor, in that it assumes that the quark and antiquark which form the meson carry equal fractions of the photon energy. It can be calculated along the same lines as the impact factor of Eq.(4.44). Nevertheless, it will suffice for the discussion of the general properties that follow.

With this impact factor, the function $V_\nu(\mathbf{q}, m)$ of Eq.(7.8) can be computed (this is not a straightforward calculation and here we quote only the result and refer to Bartels, Forshaw, Lotter & Wüsthoff (1996) for the details). One finds, in the limit of $\mathbf{q}^2 \gg m^2$, that

$$V_\nu(\mathbf{q}, m) \sim \frac{C}{\mathbf{q}^2} \ln \frac{\mathbf{q}^2}{m^2} \left(\frac{\mathbf{q}^2}{4} \right)^{-1/2 + i\nu} \qquad (7.18)$$

and we do not write explicitly the constant prefactor. Note that, modulo the logarithm, we could have anticipated this form on purely dimensional grounds.

To compute the scattering amplitude, we need to convolute ψ_A with ψ_B, where

$$\begin{aligned}
\psi_B^*(0, \mathbf{r}, \mathbf{q}) &= C \int d^2\mathbf{k} \, e^{i\mathbf{k} \cdot \mathbf{r}} \left(\frac{1}{m^2 + (\mathbf{k} - \mathbf{q}/2)^2} - \frac{1}{m^2 + \mathbf{q}^2/4} \right) \\
&= C \, 2\pi e^{-i\mathbf{q} \cdot \mathbf{r}/2} \left(K_0(mr) - \frac{2\pi \delta^2(\mathbf{r})}{\mathbf{q}^2/4 + m^2} \right). \qquad (7.19)
\end{aligned}$$

The delta function term gives zero upon convolution with ψ_A. The factor ψ_B does not spoil the dominance of the contribution from the region $\xi > 0$. As such, the angular part of the $\mathbf{r}$ integral can be approximated by 2π and we can use Eqs.(7.6), (7.14), (7.18) and (7.19) to write (again modulo an overall constant prefactor which is of no interest to us)

$$\begin{aligned}
\frac{\Im m \, A(s, t)}{s} &\sim \frac{C^2}{q^3} \ln \frac{q^2}{m^2} \frac{e^{\omega_0 y}}{(a^2 y)^{3/2}} \\
&\quad \times \int_0^{\sim 1/q} dr \, r^2 K_0(mr) \, \xi e^{-\xi^2/(4a^2 y)}. \qquad (7.20)
\end{aligned}$$

The r-integral can now be done since it is safe to take the small argument expansion of the Bessel function because $m^2 \ll q^2$, i.e. $K_0(mr) \approx \ln(1/mr)$. We integrate over the dominant region of $\xi > 0$, i.e. contributions from $r \gtrsim 1/q$ are heavily suppressed, and find

$$\frac{\Im m \, A(s,t)}{s} \sim \frac{C^2}{q^6} \ln^2\left(\frac{q^2}{m^2}\right) \frac{e^{\omega_0 y}}{(a^2 y)^{3/2}} \qquad (7.21)$$

and hence,

$$\frac{d\sigma}{dt} \sim \frac{C^4}{t^6} \ln^4\left(\frac{q^2}{m^2}\right) \frac{e^{2\omega_0 y}}{(a^2 y)^3}. \qquad (7.22)$$

A large-t elastic-scattering process that is typically more accessible to experiment is that of parton–parton elastic scattering. This has been investigated in hadron–hadron collisions (Abe *et al.* (1995), Abachi *et al.* (1994)) and in photon–hadron collisions (Derrick *et al.* (1996b)). In such processes a pair of partons (one from each 'hadron') scatter elastically off each other via the exchange of a Pomeron to produce a pair of jets which are separated by a large gap in rapidity. To lowest order, the transverse momentum of the jets produced by the scattered partons is equal to (the modulus of) the momentum transfer, $|t|$. We have chosen not to focus on these processes owing to the complications discussed at the end of Section 4.5 which arise whenever the Pomeron couples to a single parton.

So we have demonstrated that elastic scattering at large enough $-t$ can be calculated in perturbative QCD,[†] at least in those cases where the dominant contribution arises due to the exchange of a pair of (interacting) reggeized gluons. In particular, one can envisage elastic scattering processes where the dominant contribution is *not* due to Pomeron exchange. For example proton–proton elastic scattering at high-t is dominated by (at the Born level) three-gluon exchange. This is because if one views the scattering as occurring between the three constituent quarks in each proton then it is preferential to deflect each quark through the same angle. At lowest order, this then requires three gluons to be exchanged (each

---

[†] Of course the impact factors ($\Phi_{A,B}$) are generally not calculable in perturbation theory. What we have shown is that this physics essentially factorizes and the exchange dynamics is dominated by the perturbative contribution.

coupling to three constituent quarks per proton). So although one pays the price of an additional power of α_s this is more than compensated by the need to share the kick delivered by the momentum transfer equally between the constituents (Landshoff (1974)).

7.3   Diffraction dissociation and Pomeron substructure

Our focus in this section will be on the dissociation of a high Q^2 photon in $\gamma^* p \rightarrow Xp$ (where X denotes the diffracted system). This process is of particular interest since one can think of performing *deep inelastic scattering off a Pomeron target*. The possibility of unravelling the partonic substructure of the Pomeron thus presents itself. We shall have more to say on this interpretation a little later.

However, in order to prepare the ground for our discussion of the photon dissociation process we wish first to return to the inclusive deep inelastic process and its interpretation in the proton rest frame. This way of looking at the inclusive process will better equip us for our study of that subset of events containing a fast-forward proton (i.e. photon dissociation).

7.3.1   The proton rest frame picture

Recall the impact factors for deep inelastic scattering, Eq.(A.6.15) and Eq.(A.6.16). Using the identities

$$\int \frac{d^2\mathbf{l}}{(\mathbf{l}^2 + \epsilon^2)((\mathbf{l} - \mathbf{k})^2 + \epsilon^2)} = \int d^2\mathbf{r}\, e^{i\mathbf{k}\cdot\mathbf{r}}\, K_0^2(\epsilon r) \qquad (7.23)$$

and

$$\int \frac{d^2\mathbf{l}}{(\mathbf{l}^2 + \epsilon^2)^2} = \int d^2\mathbf{r}\, K_0^2(\epsilon r), \qquad (7.24)$$

we can re-write the longitudinal impact factor as follows (replacing ρ in Eq.(A.6.15) by z),

$$\begin{aligned} \Phi_L(\mathbf{k}) &= 32\alpha\alpha_s \sum_{q=1}^{n_f} e_q^2 \int_0^1 dz \int d^2\mathbf{r}(1 - e^{i\mathbf{k}\cdot\mathbf{r}}) \\ &\times Q^2\, z^2(1-z)^2\, K_0^2(\epsilon r), \end{aligned} \qquad (7.25)$$

where $\epsilon^2 = z(1-z)Q^2$.

Thus, the cross-section for the scattering of longitudinal photons is (see Eqs.(6.3) and (6.5))

$$\sigma_L(x, Q^2) = \int dz d^2\mathbf{r} \, |\Psi_L(z,r)|^2 \sigma(x,r), \qquad (7.26)$$

where

$$|\Psi_L(z,r)|^2 = \frac{6}{\pi^2}\alpha \sum_{q=1}^{n_f} e_q^2 Q^2 \, z^2(1-z)^2 \, K_0^2(\epsilon r) \qquad (7.27)$$

and

$$\sigma(x,r) = \frac{4\pi\alpha_s}{3} \int \frac{d^2\mathbf{k}}{\mathbf{k}^4} \mathcal{F}(x,\mathbf{k})(1 - e^{i\mathbf{k}\cdot\mathbf{r}}). \qquad (7.28)$$

Similarly the cross-section for the scattering of transverse photons is given by

$$\sigma_T(x, Q^2) = \int dz d^2\mathbf{r} \, |\Psi_T(z,r)|^2 \sigma(x,r), \qquad (7.29)$$

where

$$|\Psi_T(z,r)|^2 = \frac{3}{2\pi^2}\alpha \sum_{q=1}^{n_f} e_q^2 \, [z^2 + (1-z)^2]\epsilon^2 K_1^2(\epsilon r). \qquad (7.30)$$

By writing the cross-sections in such a way we have made explicit a result which has a very clear physical interpretation. In the proton rest-frame, and for low enough values of x, the photon produces the q–$\bar{q}$ pair a long distance 'down stream' of the proton (as indicated in Fig. 7.4). Some (long) time later, this pair then scatters coherently off the proton. The typical time-scale of the interaction (of the q–$\bar{q}$ pair) with the proton is very short (relative to the formation time of the pair) and as such we can consider *the transverse size of the pair to be fixed over the time of the interaction*. Consequently, we can interpret $\sigma(x,r)$ as the cross-section for the scattering of a q–$\bar{q}$ pair of transverse size r off the target proton and $\Psi(z,r)$ as the wavefunction describing the formation of a q–$\bar{q}$ pair where z and $1-z$ are the fractions of the photon energy carried by the quark and antiquark. We shall shortly justify the precise normalizations of the wavefunctions. Let us first make this physical picture a little more explicit.

We work in terms of light-cone variables, i.e. the photon momentum is written, $q = (q^+, q^-, \mathbf{0})$ where

$$q^+ = q_0 + q_3 \gg q^- = q_0 - q_3 = -Q^2/q^+$$

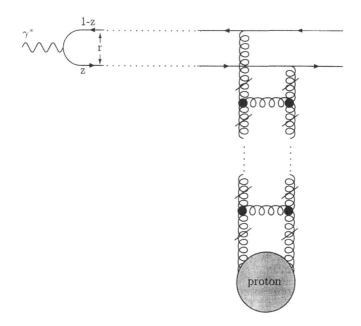

Fig. 7.4. The formation of a q-$\bar{q}$ pair from a virtual photon, followed a long time (on the scale of hadronic interactions) later by
the scattering of the pair off a proton via the exchange of a Pomeron (which, for small enough r, is the ladder of reggeized gluons
shown).

(q_μ are the components of the photon four-momentum vector).
The quark carries momentum,

$$l_q = \left(zq^+, \frac{\mathbf{l}^2}{zq^+}, \mathbf{l} \right),$$

and the antiquark momentum is obtained by replacing $z \to 1 - z$
and $\mathbf{l} \to -\mathbf{l}$. Putting the quarks on shell, we see that the energy
imbalance, ΔE, between the photon and the q-$\bar{q}$ pair, is given by

$$\Delta E = (l_q^+ + l_q^- + l_{\bar{q}}^+ + l_{\bar{q}}^- - q^+ - q^-)/2$$

$$= \frac{1}{2q^+} \left(Q^2 + \frac{\mathbf{l}^2}{z(1-z)} \right).$$

Now since $2p \cdot q = Q^2/x$ and $p = (M_p, M_p, 0)$ (M_p is the proton mass) it follows that

$$q^+ \approx \frac{Q^2}{M_p x}$$

and hence

$$\Delta E \approx \left(Q^2 + \frac{l^2}{z(1-z)} \right) \frac{M_p x}{2Q^2}.$$

Provided $l^2/(z(1-z)) \lesssim Q^2$ (which will always be the case in our subsequent considerations), it then follows, from the uncertainty principle, that the q–$\bar{q}$ pair propagate typical longitudinal distances $\sim 1/(M_p x)$ before interacting with the proton. Since we are in the low-x regime, these distances can be huge on the scale of the proton radius. Put another way, the lifetime of the q–$\bar{q}$ fluctuations of the virtual photon is huge in comparison with the typical time over which the pair interact with the target; as such we can consider the transverse size of the pair to be frozen over the time of interaction.

It is now time to make the identification of the wavefunction and cross-section more precise (in particular to determine their normalizations). Our strategy is first to establish the normalization of the transversely polarized photon wavefunction. We will then be able to infer the normalization of the cross-section $\sigma(x, r)$ and (since this cross-section does not depend upon the photon polarization) this will be sufficient to fix the normalization of the longitudinal photon wavefunction.

To lowest order, the virtual photon can either interact as a photon or via its fluctuation into a fermion–antifermion pair, i.e. denoting the physical state by $|\gamma_{\mathrm{phys}}\rangle$ we have

$$|\gamma_{\mathrm{phys}}\rangle = \sqrt{Z_3}|\gamma_B\rangle + c|f\bar{f}\rangle. \qquad (7.31)$$

Z_3 is the photon wavefunction renormalization constant, $|\gamma_B\rangle$ denotes the bare photon state and c is some coefficient (to be determined) which determines the probability that the photon is to be found in the f–$\bar{f}$ state (f labels the fermion type). Note that since we are including the possibility that the photon fluctuates into the f–$\bar{f}$ pair (i.e. $c^2 = \mathcal{O}(\alpha)$) we must work to the same order in the bare photon renormalization (i.e. $Z_3^2 = 1 + \mathcal{O}(\alpha)$).

Since we are interested in the (dominant) strong interactions of the photon with the target it follows that we are only interested in the q–$\bar{q}$ fluctuations of the virtual photon. We need to compute the coefficient, c, in this case. Working with properly normalized states, i.e. $\langle\gamma|\gamma\rangle = \langle\gamma_B|\gamma_B\rangle = \langle q\bar{q}|q\bar{q}\rangle = 1$ it follows that

$$c^2 = 1 - Z_3. \tag{7.32}$$

For transversely polarized photons, Z_3 is ultra-violet divergent. Imposing an ultra-violet cut-off Λ on the transverse momentum of the q–$\bar{q}$ pair we can write (keeping only the leading logarithm in the ultra-violet cut-off)

$$c^2 \approx \frac{\alpha}{\pi} \sum_{q=1}^{n_f} e_q^2 \ln\frac{\Lambda^2}{Q^2}. \tag{7.33}$$

(and we have summed over the three colours of quark). Since

$$\int dz d^2\mathbf{r} |\Psi_T(z,r)|^2 = c^2, \tag{7.34}$$

we have therefore fixed the normalization of the wavefunction for transverse photons. It is easy to check that this is consistent with the definitions given in Eqs.(7.26)–(7.30).

It is important to realize that Eqs.(7.26) and (7.29) are perfectly general (i.e. they are valid beyond perturbation theory). This is because they are determined purely by the space-time structure of the process. For small size q–$\bar{q}$ pairs, we can compute the wavefunction, $\Psi_{L,T}$, and the radiative corrections to $\sigma(x,r)$ which determine the QCD scaling violations. For larger sizes, perturbation theory is useless. For example, in pion–proton scattering Eq.(7.26) can be used to determine the scattering of the lowest Fock state (q–$\bar{q}$) component of the pion off the proton. In this case the pion wavefunction, $\Psi_\pi(z,r)$, is normalized to unity.

It is correct to say that by working in this representation we have succeeded in diagonalizing the scattering matrix. To see this consider the elastic-scattering amplitude, $A(s,0)$. In terms of the T-matrix elements

$$\Im m A(s,0) = \langle\gamma|T|\gamma\rangle. \tag{7.35}$$

We can expand the photon state as a sum over the interaction eigenstates, $|\psi_k\rangle$ i.e.

$$|\gamma\rangle = \sum_k c_k|\psi_k\rangle, \tag{7.36}$$

where $\sum_k |c_k|^2 = 1$. That the $|\psi_k\rangle$ are eigenstates of the interaction means that

$$T|\psi_k\rangle = p_k|\psi_k\rangle, \qquad (7.37)$$

where p_k is the probability that eigenstate k scatters off the target. So the imaginary part of the elastic scattering amplitude can be written

$$\Im m A(s,0) = \sum_k |c_k|^2 p_k. \qquad (7.38)$$

Comparing with Eqs.(7.26) and (7.29) we identify the interaction eigenstates with the set of parton states at fixed impact parameters and energy fractions (i.e. k labels the (r,z) of the interaction). p_k/s is then to be identified with the cross-section for scattering the interaction eigenstate off the target, i.e. $\sigma(r)$. Note that here we have considered the special case of elastic scattering, but it is clear that there is also the possibility of producing new states (which carry the quantum numbers of the photon). This is the process of diffraction dissociation and it is clear that the interaction eigenstates we have just been discussing are more generally the eigenstates of diffraction (of which elastic scattering is a special case). The identification of the diffraction eigenstates with the frozen partonic configurations was first made by Miettenen & Pumplin (1978).

Let us now investigate the physics of the elastic scattering amplitude. For small enough r, $\sigma(x,r) \sim r^2$, modulo scaling violations (this can be seen after expanding the exponential in Eq.(7.28) and performing the angular integral). Thus, small size pairs scatter with a cross-section which vanishes as the square of their separation. For large enough r, confinement dictates that the cross-section should saturate at a constant value of the order of a typical hadronic size. Both of these properties are necessary in order to understand the scaling of the deep inelastic structure functions (modulo the scaling violations induced by QCD corrections). Let us see why this is so.

We need to examine the Q^2-behaviour of the longitudinal and transverse cross-sections (Eqs.(7.26) and (7.29)) arising from the contributions from large size and small size q–$\bar{q}$ pairs. We expect the contributions from small size pairs to be calculable in perturbation theory whereas those from large size pairs are expected to be dominated by non-perturbative effects.

The modified Bessel functions $K_i(\epsilon r)$ in Eqs.(7.27) and (7.30) are exponentially suppressed for $\epsilon r \gg 1$. In order to extract the Q^2-behaviour it suffices to replace $K_0(\epsilon r)$ by $c_0 \Theta(1 - \epsilon r)$, i.e. a constant value, c_0, for $\epsilon r < 1$ and zero otherwise[†] and $K_1(\epsilon r)$ by $c_1 \Theta(1 - \epsilon r)/\epsilon r$.

Consider first the contribution to the cross-section which arises from large size pairs, i.e. $r \gtrsim R \gg 1/Q$ ($R \sim 1$ fm). The requirement $\epsilon r < 1$ means that the z-integration is restricted to regions near the end-points, i.e $z \lesssim 1/Q^2 R^2$ or $(1 - z) \lesssim 1/Q^2 R^2$. Thus the z-integrations give for the squared wavefunctions, $|\Psi_L(z,r)|^2$ and $|\Psi_T(z,r)|^2$,

$$\int_{\epsilon < 1/R} Q^2 z^2 (1 - z)^2 dz \sim \frac{1}{Q^4 R^6},$$

for the longitudinal cross-section and

$$\int_{\epsilon < 1/R} \left(z^2 + (1 - z)^2 \right) \frac{dz}{R^2} \sim \frac{1}{Q^2 R^4},$$

for the transverse cross-section. The integration over $\mathbf{r}$ gives (from dimensional analysis)

$$\int_R^\infty d^2\mathbf{r}\, \sigma(x,r) \sim R^2.$$

Thus the tranverse cross-section has a large size pair contribution which scales, i.e. it is proportional to $1/Q^2$, whereas the longitudinal cross-section has a large size pair contribution which is suppressed by a further factor of $1/Q^2$.

Now consider the contribution from small size pairs, i.e. $r \lesssim 1/Q$. In this case the z integration is not restricted to the end-points; $z \sim \frac{1}{2}$ and the quark–antiquark pair share the photon energy roughly equally. The contributions from the z-integrations for the longitudinal and transverse cross-sections are both proportional to Q^2. If the scattering cross-section, $\sigma(x,r)$, is calculated in leading order in perturbation theory (i.e. two-gluon exchange rather than the complete Pomeron ladder shown in Fig. 7.4) then on dimensional grounds we have for the integration over $\mathbf{r}$

$$\int_0^{1/Q} d^2\mathbf{r}\, \sigma(x,r) \sim \frac{1}{Q^4},$$

[†] As $r \to 0$ the function $K_0(\epsilon r)$ behaves like $\log r$. However, this is an integrable singularity and does not affect the validity of this approximation.

so that both the longitudinal and transverse cross-sections scale. Inclusion of the complete ladder (QCD Pomeron) in the calculation of $\sigma(x, r)$ leads to the scaling violations discussed in the preceding chapter.

An alternative way of seeing these same results is to use the fact that, from purely dimensional grounds, we can write

$$\int d^2\mathbf{r}\, \sigma(r) K_i^2(\epsilon r) \propto \frac{1}{\epsilon^4}. \tag{7.39}$$

This is kept finite since $\epsilon^2 \sim m_q^2$ as $z \to 0, 1$. Here the quark mass, m_q, acts as the confining scale. So,

$$\sigma_T \sim \int dz \frac{z^2 + (1 - z)^2}{\epsilon^2}$$

and

$$\sigma_L \sim \int dz \frac{Q^2[z(1 - z)]^2}{\epsilon^4}.$$

It is clear that the end-point contribution to the z-integral leads to the $1/Q^2$ behaviour of σ_T and the m_q^2/Q^4 behaviour of σ_L. Also, the $z \sim \frac{1}{2}$ contributions clearly yield the $1/Q^2$ behaviour for both longitudinal and transverse cross-sections.

Thus, we have regained the property of Bjorken scaling (neglecting the QCD corrections contained in $\sigma(x, r)$). However, we have gained a little more insight into the final state morphology of low-x deep inelastic events. There are large contributions to the cross-section for scattering of transverse photons from the so-called **aligned jet** configurations (where one parton carries all the photon energy). The small-size configurations also generate a scaling contribution and are associated with the more democratic final state in which the quark and antiquark share the photon energy. The scaling violations to the structure function are also calculable in perturbation theory; only the small size fluctuations evolve in Q^2. Furthermore, since the longitudinal cross-section is determined by small size fluctuations (the large size fluctuations being suppressed by an extra power of $1/Q^2$) we are able to write $F_L(x, Q^2)$ directly in terms of the parton densities evaluated at the scale, Q^2. For the transverse cross-section the scaling non-perturbative part arising from large size fluctuations must be obtained from experiment at some fixed Q^2, whereas the Q^2-evolution can be calculated perturbatively.

7.3.2 Introduction to photon dissociation

We have spent most of this book talking about elastic scattering and, through the optical theorem, total cross-sections. In the preceding subsection we highlighted the fact that the existence of elastic scattering naturally suggests the possibility of diffraction dissociation. A beam of hadrons scattered off some target will typically be either absorbed (perhaps leading to the excitation of the target or emission of some final state particles), scattered elastically or diffracted. What is the physical picture which underpins the connection between the total cross-section, elastic scattering and diffraction scattering? The answer, not surprisingly, lies in an analogy with the physics of diffraction in wave optics. Before discussing the special case of photon dissociation, we wish to spend some time making clear the connection between these apparently such different processes.

Consider a broad beam of plane polarized light, incident on some small piece of polaroid (the target). If the light is polarized at some non-zero angle (relative to the axis of the polaroid) then the component that is polarized parallel to the axis of the polaroid will pass through without scattering, i.e. for this component it is as though the polaroid were absent. The other component, which has its axis of polarization perpendicular to the axis of the polaroid, has a small section of its wavefront which is totally absorbed on passing the polaroid, so that the wavefront is partitioned into two wavefronts which pass either side of the polaroid and interfere with each other producing a diffraction pattern. This diffraction pattern is detected (over and above the constant background from the unscattered component) some distance behind the polaroid. Since the diffracted wave is polarized normal to the axis of the polaroid it necessarily contains a component which is polarized parallel to the (polarization of the) incoming wave and also a component which is polarized perpendicular to the incoming wave.

What has this to do with, for example, scattering a beam of hadrons off some target (e.g. another hadron or a nucleus)? The absorption of the light beam in the polaroid is analogous to the inelastic scattering of the hadron on the target (e.g. producing an excited nuclear state or some multi-particle final state). The diffraction of the incoming wave into the component which carries the same polarization is analogous to the elastic scattering of the beam particle. Finally, we saw that diffraction can lead to the production of a new state (carrying polarization distinct from the incoming beam); this, too, should have an analogy in particle physics: this is what we call diffraction dissociation. New states can be 'diffracted into existence' by the interaction with the target. We say that *the diffractive processes are the shadow of the inelastic processes.*

More discussion of the physical picture can be found in the paper by Good & Walker (1960), where beam diffraction was first considered. Let us merely note that in order to open up the diffractive channel, it is important to have energy degeneracies (up to some approximation). In the optical case discussed above the two polarization states were degenerate in energy. In the particle physics case the effective degeneracy is achieved by working at high centre-of-mass energies (so that all masses are small relative to the centre-of-mass energy and the proton does not dissociate). This is why diffraction is characterized by processes which involve large gaps in rapidity.

We are now able to commence our study of the rapidity gap events in deep inelastic scattering. We will start by looking at the simplest diffracted system, namely, the one in which the photon dissociates into a single q–$\bar{q}$ pair, which is separated from the fast moving final-state proton by a large gap in rapidity. A typical contribution to the amplitude is shown in Fig. 7.5. We will work in the so-called Born approximation, i.e. the exchange is modelled by the exchange of two gluons (the BFKL corrections will not alter our essential conclusions). In this case, the cross-section for scattering the q–$\bar{q}$ colour dipole off the proton is only a function of the dipole size, r. Notice that the momentum transfer t is no longer zero; in fact simple kinematics allows us to show that, for $M_p^2 \ll Q^2, M_X^2 \ll W^2$,

$$-t_{\min} = (M_X^2 + Q^2)^2 \frac{M_p^2}{W^4}, \qquad (7.40)$$

where M_X is the invariant mass of the diffractive system (in this case the q–$\bar{q}$ pair) and W is the γ–p centre-of-mass energy. Clearly, for large enough W, $t_{\min}$ is very small (on the scale of the hadron mass). Since we insist that the proton remain intact, it follows

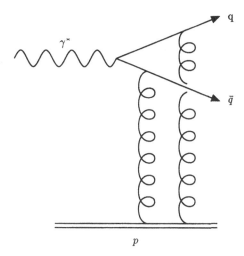

Fig. 7.5. One of the contributions to the amplitude for the process $\gamma p \rightarrow q\bar{q}p$.

that $-t \sim -t_{\min}$ since larger values of $-t$ are suppressed by some hadronic form factor (i.e. it is unlikely that the proton will remain intact after undergoing a large t interaction). Consequently, we will assume that $t = 0$ is a good approximation.

Due to the space-time picture which was discussed in the preceding subsection, we are able to think of the q–$\bar{q}$ pair in terms of eigenstates of the diffraction scattering matrix. Consequently, we can write the amplitude as a convolution of the squared amplitude for the $\gamma \rightarrow q\bar{q}$ formation with the square of the dipole cross-section, i.e.

$$\frac{d\sigma_{T,L}^D}{dt}\bigg|_{t=0} = \int_0^1 dz \int d^2\mathbf{r} |\psi_{T,L}(z,r)|^2 \frac{\sigma(r)^2}{16\pi}. \qquad (7.41)$$

Let us start by proving this result. In the notation of the preceding subsection, we can write

$$\frac{d\sigma^D}{dt}\bigg|_{t=0} = \frac{\sum_k |\langle \gamma|T|\psi_k\rangle|^2}{16\pi s^2} - \frac{d\sigma^{el}}{dt}\bigg|_{t=0} \qquad (7.42)$$

and we have subtracted off the elastic cross-section in order to define the diffractive rate. Substituting for the elastic rate gives

$$\left.\frac{d\sigma^D}{dt}\right|_{t=0} = \frac{1}{16\pi s^2}\left(\sum_k |\langle\gamma|T|\psi_k\rangle|^2 - |\langle\gamma|T|\gamma\rangle|^2\right). \qquad (7.43)$$

Decomposing the photon into the sum over scattering eigenstates gives

$$\left.\frac{d\sigma^D}{dt}\right|_{t=0} = \frac{1}{16\pi s^2}\left[\sum_k |c_k|^2 p_k^2 - \left(\sum_k |c_k|^2 p_k\right)^2\right]. \qquad (7.44)$$

Identifying the label k with the pair size and longitudinal momentum fraction we thus arrive at our final result:

$$\left.\frac{d\sigma^D}{dt}\right|_{t=0} = \frac{\langle\sigma^2\rangle - \langle\sigma\rangle^2}{16\pi}, \qquad (7.45)$$

where we have written the cross-section averaged over the photon wavefunction in a compact form, i.e.

$$\langle\sigma^n\rangle \equiv \int dz d^2\mathbf{r}|\psi(z,r)|^2\sigma(r)^n. \qquad (7.46)$$

Neglecting the second term (since it is suppressed by a power of α for photon scattering) we thus establish the validity of Eq.(7.41).

The essential difference in comparison with the inclusive case is the presence of the extra factor of $\sigma(r)$. By arguments along the lines of those of the preceding subsection, we now see that diffractive $q-\bar{q}$ production from transverse photons is dominated by large sizes of the $q-\bar{q}$ pair, i.e. the aligned jet configurations (Bjorken (1994)). Note also that the leading behaviour is $\sim 1/Q^2$ (i.e. the same order in Q^2 as the inclusive cross-section). Contrast this with the inclusive cross-section, where the leading (scaling) contribution samples both large and small size pairs. In the diffractive case, the extra factor of $\sigma(r)$ leads to the suppression of the short-distance contribution by a power of $1/Q^2$. For the production from longitudinal photons, we have the result that the short and long distance contributions mix. However, both contributions are suppressed by a power of Q^2 relative to the rate for production from transverse photons.

Note that, in those cases where the large size configurations dominate, it is no longer a good approximation to neglect the quark mass contributions (this is because $\epsilon^2 \sim m_q^2$). Moreover, for the large size configurations we have no right to use perturbation theory and must (in the absence of any fundamental theory of the non-perturbative regime) rely on a more phenomenological approach. *Despite the fact that the photon has a large virtuality, we have shown that the dominant contribution to the photon dissociation process is of non-perturbative origin.*

Even so, we can go a little further. We can derive an approximate expression for the dependence of the cross-section on the diffracted mass, M_X. If the quark has four-momentum, l_q^μ, where

$$l_q^\mu = (E_q, zp_\gamma, \mathbf{l}),$$

then $l_q^2 = m_q^2$ fixes

$$E_q \approx zp_\gamma + \frac{\mathbf{l}^2 + m_q^2}{2zp_\gamma}$$

in terms of the photon momentum, p_γ. The antiquark four-momentum is once again obtained by replacing $z \to 1 - z$ and $\mathbf{l} \to -\mathbf{l}$. The diffracted mass is defined to be the invariant mass of the diffracted system, i.e.

$$M_X^2 \equiv (l_q + l_{\bar q})^2 \approx \frac{\mathbf{l}^2 + m_q^2}{z(1 - z)}. \tag{7.47}$$

For the non-perturbative (large size) configurations (which dominate the diffractive rate), the z-integral is dominated by the regions of z close to 0 or 1. From the z near zero region we have,

$$M_X^2 \approx \frac{m_q^2}{z}$$

(since the large size pairs have $\mathbf{l}^2 \sim 0$). Hence we can undo the z-integral and write

$$\frac{d\sigma_T^D}{dt dM_X^2} \sim \frac{m_q^2}{M_X^4} \int d^2\mathbf{r} |\Psi_T(z, r)|^2 \sigma(r)^2. \tag{7.48}$$

As in Eq.(7.39) we see that (from dimensional analysis)

$$\int d^2\mathbf{r}\,\sigma(r)^2 \epsilon^2 K_i^2(\epsilon r) \propto \frac{1}{\epsilon^4}$$

and, since

$$\begin{aligned} \epsilon^2 &= Q^2 z(1 - z) + m_q^2 \\ &\approx m_q^2 \left(1 + \frac{Q^2}{M_X^2}\right), \end{aligned} \tag{7.49}$$

we have

$$\frac{d\sigma_T^D}{dt\,dM_X^2} \sim \frac{1}{m_q^2}\frac{1}{(Q^2+M_X^2)^2}. \tag{7.50}$$

The contribution from the region $z \sim 1$ of course yields the same form. This expression will be a good approximation provided $M_X^2 \gtrsim Q^2 \gg m_q^2$ since for small M_X^2 we can no longer assume that the dominant contribution arises from the end points of the z-integral.

The rate for production of large diffracted masses falls away as $\sim 1/M_X^4$ at large M_X^2. The origin of this strong decrease can be traced back to the fact that the q–$\bar q$ pair scatters directly off the target. It is also possible to radiate additional gluons off the original q–$\bar q$ pair and then scatter the resulting multi-parton configuration (frozen in impact parameter) off the target. Of course the radiation of more partons occurs at the price of additional powers in the strong coupling. However, the spin-1 nature of the gluon ensures a weaker decay at large M_X^2. In fact one expects a $\sim 1/M_X^2$ behaviour. We do not pursue these details at this stage. In the next chapter we will consider the higher Fock components of the photon wavefunction.

7.3.3 The Pomeron structure function

In a frame in which the proton is fast-moving, it is tempting to think of the photon as probing the structure of a Pomeron which has been offered up as an effective target by the proton. The picture (shown in Fig. 7.6) suggests the following form for the diffractive cross-section:

$$\frac{d\sigma_T^D}{dt\,dx_P} = \frac{4\pi^2\alpha}{Q^2}f(x_P)F_T^P(\beta,Q^2), \tag{7.51}$$

where x_P is the fraction of the incoming proton energy which is carried away by the Pomeron and β is the fraction of the Pomeron momentum carried by the struck quark. The Bjorken-x $(= Q^2/2p \cdot q)$ is therefore the product βx_P.

$F_T^P(\beta, Q^2)$ is the structure function of the Pomeron for scattering off transverse photons (it scales in the absence of QCD corrections) and $f(x_P)$ is a factor which determines the Pomeron flux. For simplicity we ignore any t-dependence on the right hand

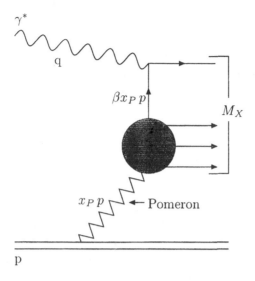

Fig. 7.6. Diffractive photon dissociation process in which a Pomeron is 'emitted' by the proton with fraction, x_P, of the proton momentum, p. The Pomeron is probed by a virtual photon of momentum q, which strikes a quark inside the Pomeron that carries a fraction β of the Pomeron momentum. M_X is the invariant mass of the Pomeron–photon system.

side (i.e. we are interested only in the behaviour near $t = 0$). Since $(x_P p + q)^2 = M_X^2$ and $W^2 = (p + q)^2$ (p is the proton four-momentum) we have

$$x_P = \frac{Q^2 + M_X^2}{Q^2 + W^2},\qquad (7.52)$$

and

$$\beta = \frac{x}{x_P} = \frac{Q^2}{2x_P \, p \cdot q} = \frac{Q^2}{Q^2 + M_X^2}.\qquad (7.53)$$

In terms of these variables, we can re-write the diffractive cross-section for q–$\bar{q}$ production (see Eq.(7.50)) as

$$\frac{d\sigma_T^D}{dx_P} \sim \frac{\beta}{x_P}\frac{1}{Q^2}\qquad (7.54)$$

and hence

$$F_T^P(\beta, Q^2) \sim \beta,$$

$$f(x_P) \sim \frac{1}{x_P}. \tag{7.55}$$

So the concept of the Pomeron structure function makes sense, at least in the approximation that the diffracted system is a pure q-$\bar{q}$ pair of not too small invariant mass.

Note that for $M_X^2 \leq 4m_q^2$ the cross-section must vanish (M_X must be larger than the mass of the quark plus antiquark). For $m_q^2 \ll Q^2$ this means that $\beta \approx 1$. We can crudely account for this effect by taking a Pomeron structure function of

$$F_T^P(\beta, Q^2) \sim \beta(1 - \beta). \tag{7.56}$$

This modification is consistent with Eq.(7.55) since it corresponds to multiplication by the factor $M_X^2/(Q^2 + M_X^2)$ which is ~ 1 in the region we are considering. Note that in reality the suppression as $M_X^2 \to 4m_q^2$ is faster than any power of $1 - \beta$. To see this, notice that $M_X^2 \to 4m_q^2$ corresponds to $z \to \frac{1}{2}$ and $\mathbf{l}^2 \to 0$ (i.e. $r \gg 1/Q$). This is the region where the argument of the Bessel function is large and leads to an exponential suppression.

Processes which probe the structure of the Pomeron (at $t = 0$) are termed **hard-diffractive**. The two main types of process we have in mind are deep inelastic diffraction (discussed above) and those processes where hadronic jets are produced in the diffracted system, as in Fig. 7.7. Of course these are the analogous processes to their non-diffractive counterparts, which are used to constrain hadronic parton densities.

If the concept of a Pomeron structure function is to be useful it should be *universal*. That is to say there should exist hard-diffractive processes which are driven by a common set of Pomeron parton distribution functions. This property of universality is certainly not an obvious consequence of QCD. In those hard-diffractive processes where soft physics dominates and the soft Pomeron (which is, so far, well described as a simple Regge pole) is exchanged we expect the universality of Pomeron parton distribution functions to apply. However, if the soft Pomeron pole is not the dominant exchange then we, a priori, have no good reason to expect the factorization of the x_P- and β-dependence (and even if factorization does hold there is no reason to expect universality of the extracted parton densities). Let us briefly explain how the Regge model leads to factorizable and universal Pomeron parton

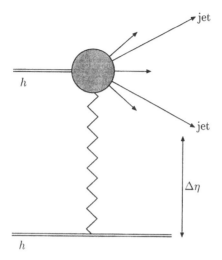

Fig. 7.7. Hard-diffractive production of jets in hadron–hadron diffractive scattering. The zig–zag line denotes the Pomeron exchange and $\Delta\eta$ denotes the final state rapidity gap.

distribution functions.

For the general diffractive process, $A + B \rightarrow A + X$, and assuming the Pomeron to be a simple Regge pole, we can write the diffractive cross-section as

$$M_X^2 \frac{d\sigma}{dt dM_X^2} = \frac{1}{16\pi} |\beta_A(t)|^2 \left(\frac{s}{M_X^2}\right)^{2\alpha_P(t)-2} \sigma_{BP}(M_X^2, t), \quad (7.57)$$

where $\beta_A(t)$ reflects the coupling of the Pomeron to the target, A, and σ_{BP} is the total cross-section for scattering particle B off the Pomeron.

In the case where the beam particle is a virtual photon, we probe the quark densities of the Pomeron, i.e. for scattering of transverse and longitudinal photons

$$\sigma_{\gamma P} \equiv \frac{4\pi^2\alpha}{Q^2} F_{T,L}^P(\beta, Q^2). \quad (7.58)$$

The Pomeron quark densities are defined by

$$F_2^P(\beta, Q^2) = F_L^P(\beta, Q^2) + F_T^P(\beta, Q^2) = \sum_i e_i^2 x f_{q/P}(\beta, Q^2).$$

These parton densities can also be probed in, for example, the hard-diffractive jet production process of Fig. 7.7 in which case we can write,

$$\frac{d\sigma(A + B \rightarrow A + jjX)}{dt\,dM_X^2\,dp_T^2} = \frac{1}{16\pi}|\beta_A(t)|^2 \left(\frac{s}{M_X^2}\right)^{2\alpha_P(t)-2}$$
$$\times \frac{d\sigma(BP \rightarrow jjX)}{dp_T^2}, \qquad (7.59)$$

where
$$\frac{d\sigma(BP \rightarrow jjX)}{dp_T^2} = \sum_{i,j} \int dx_1 f_{i/B}(x_1, p_T^2) \int dx_2 f_{j/P}(x_2, p_T^2)$$
$$\times \frac{d\hat{\sigma}(ij)}{dp_T^2}. \qquad (7.60)$$

The sum is over all parton types, x_1 is the fraction of particle B's momentum carried by parton i and x_2 is the fraction of the Pomeron momentum carried by parton j. The differential cross-section, $d\hat{\sigma}/dp_T^2$, is that of the hard sub-process, i.e. the scattering of partons i and j into the final state (producing a pair of partons with transverse momentum p_T relative to the collision axis) and is straightforward to compute in perturbative QCD.

Thus we see that Pomeron parton densities can be extracted from data on hard-diffractive processes just as proton parton densities can be extracted from hard non-diffractive scattering. In the case of the proton parton densities, one has the advantage of a momentum sum rule, which allows a constraint to be placed upon the size of the gluon density from a measurement of the quarks. It is far from clear that a similar sum rule holds for the Pomeron.

Of course the Pomeron is much more elusive than a hadron. Indeed, there is some ambiguity in using a single word to describe a wide range of phenomena. It remains to be seen whether the object which drives hard-diffractive jet production is the same as that which drives the rapidity gap processes in deep inelastic scattering and even whether the Regge inspired picture of the Pomeron as an effective target 'particle' is valid.

7.4 Summary

- Keeping the four-momentum transfer to the BFKL Pomeron

large is an excellent way to ensure the dominance of perturbative dynamics. The momentum transfer, $-t = \mathbf{q}^2$, acts as an effective infra-red cut-off. Contributions from Pomeron sizes larger than $\sim 1/q$ are heavily suppressed whilst the dominant contributions (from sizes $\gtrsim 1/q$) are characterized as a solution to the diffusion equation.

• In the target rest frame, the high energy scattering matrix is diagonalized by eigenstates of partonic configurations whose impact parameters are frozen over the time of the interaction. This facilitates an elegant physical picture of elastic scattering and diffraction dissociation processes.

• Despite the large virtuality, dissociation of virtual photons at high energies is dominated by non-perturbative physics. This is because the dominant configurations are of an aligned-jet nature.

• It may be useful to think of the photon dissociation process as one which performs deep inelastic scattering off a Pomeron target. For not too large diffracted masses, the Pomeron structure function for scattering off transverse photons can be approximated by $\sim \beta(1 - \beta)$, where β is the Bjorken-x of the Pomeron–photon system.

8
Taming the growth

So far, this book has been concerned with the behaviour of QCD in the leading logarithmic approximation, which should be appropriate for large enough centre-of-mass energies and for those processes which satisfy the criteria relevant for the use of perturbative QCD. In Chapters 2–4, we derived and solved the BFKL equation. We were led to think of the Pomeron as the t-channel exchange of a pair of (interacting) reggeized gluons. In this chapter, we start off by reformulating the results already obtained for the elastic-scattering amplitude of two colourless states[†] in a way which suggests that we view the scattering as the incoherent scattering of individual colour dipoles whose locations in configuration space are frozen over the time of interaction. This approach will lead us to a very tangible physical picture of high energy scattering in configuration space.

In Section 8.2 we turn our attention to the undesirable feature which afflicts the scattering amplitudes calculated in the leading logarithm approximation. This is the violation of unitarity which results from the strong growth of the total cross-section with increasing energy. The dipole formalism discussed in Section 8.1 provides a very elegant framework in which to investigate the dominant corrections to the leading logarithm approximation which ensure that the theory remains unitary. We begin Section 8.2 by setting up an operator formalism (due to Mueller (1995)) to describe the dipole evolution and interaction. This formalism is subsequently used to incorporate the corrections which arise due to multiple dipole scattering effects (or, equivalently, the exchange of more than one Pomeron between the colliding hadrons) and ensure the unitarity of the scattering amplitude.

[†] We actually consider scattering of states whose leading Fock component is a heavy quark–antiquark pair, although our investigation is in principle much more general.

8.1 Dipole scattering

Let us start by defining, analogously to Eq.(4.46), the universal BFKL amplitude in impact parameter space, i.e.

$$\hat{f}(y, \mathbf{b_1}, \mathbf{b_1'}, \mathbf{b_2}, \mathbf{b_2'}) = \frac{1}{(2\pi)^4} \partial_{\mathbf{b_1}}^2 \partial_{\mathbf{b_1'}}^2 \int d^2\mathbf{k_1} d^2\mathbf{k_2} d^2\mathbf{q}$$

$$\times \; e^{i(\mathbf{k_1}\cdot\mathbf{b_{11'}} - \mathbf{k_2}\cdot\mathbf{b_{22'}} + \mathbf{q}\cdot(\mathbf{b_1'} - \mathbf{b_2'}))} \frac{F(s, \mathbf{k_1}, \mathbf{k_2}, \mathbf{q})}{\mathbf{k_2}^2(\mathbf{k_1} - \mathbf{q})^2}$$

$$= \frac{1}{(2\pi)^4} \int d^2\mathbf{k_1} d^2\mathbf{k_2} d^2\mathbf{q}$$

$$\times \; e^{i(\mathbf{k_1}\cdot\mathbf{b_{11'}} - \mathbf{k_2}\cdot\mathbf{b_{22'}} + \mathbf{q}\cdot(\mathbf{b_1'} - \mathbf{b_2'}))} \frac{\mathbf{k_1}^2}{\mathbf{k_2}^2} F(s, \mathbf{k_1}, \mathbf{k_2}, \mathbf{q}), \quad (8.1)$$

where $F(s, \mathbf{k_1}, \mathbf{k_2}, \mathbf{q})$ is the usual BFKL amplitude which determines the scattering of two gluons, of transverse momenta $\mathbf{k_1}$ and $\mathbf{k_2}$ respectively, at energy s and with momentum transfer $\mathbf{q}$. It is obtained from $f(\omega, \mathbf{k_1}, \mathbf{k_2}, \mathbf{q})$ after inverting the ω-plane Mellin transform to reveal explicitly the energy, s, dependence. We use the notation where $y \equiv \ln(s/k^2)$ and $\mathbf{b_{11'}} \equiv \mathbf{b_1} - \mathbf{b_1'}$ (and similarly for $\mathbf{b_{22'}}$).

Using Eq.(4.52) and keeping only the $n = 0$ term in the sum over n we can write

$$\hat{f}(y, \mathbf{b_1}, \mathbf{b_1'}, \mathbf{b_2}, \mathbf{b_2'}) = \frac{1}{(2\pi)^4} \int_{-\infty}^{\infty} d\nu \int d^2\mathbf{c} \frac{\nu^2}{(\nu^2 + 1/4)^2} e^{\bar{\alpha}_s \chi_0(\nu) y}$$

$$\times \; \partial_{\mathbf{b_1}}^2 \partial_{\mathbf{b_1'}}^2 \; \tilde{\phi}_0^\nu(\mathbf{b_1}, \mathbf{b_1'}, \mathbf{c}) \tilde{\phi}_0^{\nu*}(\mathbf{b_2}, \mathbf{b_2'}, \mathbf{c}). \quad (8.2)$$

Equation (4.51) then allows us to write

$$\hat{f}(y, \mathbf{b_1}, \mathbf{b_1'}, \mathbf{b_2}, \mathbf{b_2'}) = \frac{1}{\pi^4} \int_{-\infty}^{\infty} d\nu \int d^2\mathbf{c} \frac{\nu^2}{\mathbf{b_{11'}}^4} e^{\bar{\alpha}_s \chi_0(\nu) y}$$

$$\times \; \tilde{\phi}_0^\nu(\mathbf{b_1}, \mathbf{b_1'}, \mathbf{c}) \tilde{\phi}_0^{\nu*}(\mathbf{b_2}, \mathbf{b_2'}, \mathbf{c}). \quad (8.3)$$

Now let us consider the convolution

$$\int d^2\mathbf{b_x} d^2\mathbf{b_x'} \; \hat{f}(y - y', \mathbf{b_1}, \mathbf{b_1'}, \mathbf{b_x}, \mathbf{b_x'}) \hat{f}(y', \mathbf{b_x}, \mathbf{b_x'}, \mathbf{b_2}, \mathbf{b_2'}).$$

Using the results (which we quote without proof and for details we refer to Lipatov (1986))

$$\int \frac{d^2\mathbf{b_x} d^2\mathbf{b'_x}}{\mathbf{b_{xx'}}^4} \tilde{\phi}_0^\nu(\mathbf{b_x}, \mathbf{b'_x}, \mathbf{c}) \tilde{\phi}_0^{\mu*}(\mathbf{b_x}, \mathbf{b'_x}, \mathbf{c'})$$

$$= \frac{\pi^4}{2\nu^2} \delta(\nu - \mu) \delta^2(\mathbf{c} - \mathbf{c'}) + \frac{2^{4i\nu} i\pi^3}{\nu} \frac{\delta(\nu + \mu)}{|\mathbf{c} - \mathbf{c'}|^{2+4i\nu}}, \quad (8.4)$$

and

$$\int d^2\mathbf{c} \frac{\tilde{\phi}_0^{\nu*}(\mathbf{b_1}, \mathbf{b'_1}, \mathbf{c})}{|\mathbf{c} - \mathbf{c'}|^{2+4i\nu}} = \frac{\pi}{2i\nu} 2^{-4i\nu} \tilde{\phi}_0^\nu(\mathbf{b_1}, \mathbf{b'_1}, \mathbf{c'}), \quad (8.5)$$

one can derive the important result

$$\int d^2\mathbf{b_x} d^2\mathbf{b'_x} \, \hat{f}(y - y', \mathbf{b_1}, \mathbf{b'_1}, \mathbf{b_x}, \mathbf{b'_x}) \hat{f}(y', \mathbf{b_x}, \mathbf{b'_x}, \mathbf{b_2}, \mathbf{b'_2})$$

$$= \hat{f}(y, \mathbf{b_1}, \mathbf{b'_1}, \mathbf{b_2}, \mathbf{b'_2}). \quad (8.6)$$

This is analogous to the $t = 0$ convolution of Eq.(5.2) and, as in that case, is true for arbitrary y'.

We can use Eq.(8.6) to factorize the BFKL amplitude in such a way that it can be absorbed into the definitions of the external impact factors. In particular, we can show (the details are included in Appendix A to this chapter) that

$$\frac{F(s, \mathbf{k_1}, \mathbf{k_2}, \mathbf{q})}{\mathbf{k_2^2}(\mathbf{k_1} - \mathbf{q})^2} = \frac{1}{(2\pi)^6} \int d^2\mathbf{b_{11'}} d^2\mathbf{b_{22'}} \frac{d^2\mathbf{b_{xx'}}}{\mathbf{b_{xx'}}^2} \frac{d^2\mathbf{b_{yy'}}}{\mathbf{b_{yy'}}^2}$$

$$\times \frac{1}{4} \int \frac{d^2\mathbf{l}}{\mathbf{l^2}(\mathbf{l} - \mathbf{q})^2} N(b_{11'}, b_{xx'}, y', q) \, N(b_{22'}, b_{yy'}, y - y', q)$$

$$\times e^{-i(\mathbf{k_1} - \mathbf{q}/2) \cdot \mathbf{b_{11'}}} e^{-(\mathbf{k_2} - \mathbf{q}/2) \cdot \mathbf{b_{22'}}}$$

$$\times \left[e^{i(\mathbf{q} - \mathbf{l}) \cdot \mathbf{b_{xx'}}/2} - e^{-i(\mathbf{q} - \mathbf{l}) \cdot \mathbf{b_{xx'}}/2} \right] \left[e^{i(\mathbf{q} - \mathbf{l}) \cdot \mathbf{b_{yy'}}/2} - e^{-i(\mathbf{q} - \mathbf{l}) \cdot \mathbf{b_{yy'}}/2} \right]$$

$$\times \left[e^{i\mathbf{l} \cdot \mathbf{b_{xx'}}/2} - e^{-i\mathbf{l} \cdot \mathbf{b_{xx'}}/2} \right] \left[e^{i\mathbf{l} \cdot \mathbf{b_{yy'}}/2} - e^{-i\mathbf{l} \cdot \mathbf{b_{yy'}}/2} \right]. \quad (8.7)$$

Again, non-boldface is used to denote the modulus of a two-vector. The **dipole number density** is defined by

$$N(r_0, r, y, q) \equiv \int_{-\infty}^{\infty} \frac{d\nu}{2\pi} V_q^{\nu*}(r_0) V_q^\nu(r) e^{\bar{\alpha}_s \chi_0(\nu) y} \frac{r_0}{r}, \quad (8.8)$$

where

$$V_q^\nu(r) \equiv \frac{2i\nu}{\pi r} \int d^2\mathbf{R} e^{i\mathbf{q} \cdot \mathbf{R}} \left[\frac{r^2}{(\mathbf{R} + \mathbf{r}/2)^2 (\mathbf{R} - \mathbf{r}/2)^2} \right]^{1/2+i\nu}. \quad (8.9)$$

It will soon become clear why we referred to $N(r_0, r, y, q)$ as the dipole number density, for now we take Eq.(8.8) as a definition of N.

To compute physical scattering amplitudes we need to perform the convolution with the appropriate impact factors, e.g. as in Eq.(4.36). For simplicity let us suppose that the external particles each contain only a single quark–antiquark pair, e.g. as would be the case for elastic photon–photon scattering.

In Chapter 7, we showed that the photon impact factor for $t = 0$ can be written in terms of the (light-cone) wavefunction of the photon, e.g. see Eqs.(7.25) and (7.27). In particular we derived the relation

$$\Phi(\mathbf{k}) = \frac{16\pi^2 \alpha_s}{3} \int_0^1 dz \int d^2\mathbf{r} |\Psi(z, r)|^2 (1 - e^{i\mathbf{k}\cdot\mathbf{r}}). \qquad (8.10)$$

The wavefunction, $\Psi(z, r)$, specifies the probability that the photon has fluctuated into the q–$\bar{q}$ pair of transverse size r and with their momentum partitioned in the ratio $z : (1 - z)$.

Equation (8.10) is quite general. By this we mean that for any impact factor, which describes the interaction of a q–$\bar{q}$ pair with the two gluons of the Pomeron, we can always write down the corresponding wavefunction and factorize off the factor $(1 - e^{i\mathbf{k}\cdot\mathbf{r}})$. We shall subsequently refer to the generic q–$\bar{q}$ system as an **onium** state. Let us recall the origin of the $(1 - e^{i\mathbf{k}\cdot\mathbf{r}})$ factor in Eq.(8.10). From Eqs.(7.23), (7.24) and (7.25) we see that the factor of unity arises from those two graphs where the gluons couple to the same quark (or antiquark) in the onium. The second, exponential, factor derives from the coupling to both the quark and antiquark of the onium. The cancellation between these two types of graph, which occurs whenever one of the two gluons goes on shell (and hence ensures the finiteness of the scattering amplitude), has been explicitly displayed in this factor. The above discussion was specific to the case of zero momentum transfer (i.e. $\mathbf{q} = \mathbf{0}$). For non-zero momentum transfer we have the following general relation between the impact factor and the onium wavefunction:

$$\Phi(\mathbf{k}) = \frac{8\pi^2 \alpha_s}{3} \int_0^1 dz \int d^2\mathbf{r} |\Psi(z, r)|^2$$
$$\times (e^{i\mathbf{k}\cdot\mathbf{r}/2} - e^{-i\mathbf{k}\cdot\mathbf{r}/2})(e^{i(\mathbf{q}-\mathbf{k})\cdot\mathbf{r}/2} - e^{-i(\mathbf{q}-\mathbf{k})\cdot\mathbf{r}/2}). \qquad (8.11)$$

To re-iterate, by working in the co-ordinate space representation the formation of the onium state is determined by the wavefunction factor $|\Psi(z,r)|^2$ and this can be cleanly factorized from the coupling of the q–$\bar{q}$ pair to the Pomeron (contained in the exponential factors). This is consistent with the space-time picture presented in the preceding chapter (see Section 7.3). We say that the dependence on the onium wavefunction factorizes from the coupling of the dipole (q–$\bar{q}$ pair of fixed size r) to the two gluons of the Pomeron.

The elastic-scattering amplitude of two onium states, with wavefunctions $\Psi_1(z_1,r_1)$ and $\Psi_2(z_2,r_1)$, respectively, can now be written:

$$
\frac{\Im m A(s,t)}{s} = \frac{8}{9} \int dz_1 dz_2 \int d^2r_1 d^2r_2 |\Psi_1(z_1,r_1)|^2 |\Psi_2(z_2,r_2)|^2
$$
$$
\times \alpha_s^2 \int \frac{d^2\mathbf{b_{xx'}}}{2\pi\mathbf{b_{xx'}}^2} \frac{d^2\mathbf{b_{yy'}}}{2\pi\mathbf{b_{yy'}}^2} \int \frac{d^2\mathbf{l}}{\mathbf{l}^2(\mathbf{l}-\mathbf{q})^2}
$$
$$
\times N(r_1, b_{xx'}, y', q)\, N(r_2, b_{yy'}, y-y', q)
$$
$$
\times \left[e^{i(\mathbf{q}-\mathbf{l})\cdot\mathbf{b_{xx'}}/2} - e^{-i(\mathbf{q}-\mathbf{l})\cdot\mathbf{b_{xx'}}/2} \right] \left[e^{i(\mathbf{q}-\mathbf{l})\cdot\mathbf{b_{yy'}}/2} - e^{-i(\mathbf{q}-\mathbf{l})\cdot\mathbf{b_{yy'}}/2} \right]
$$
$$
\times \left[e^{i\mathbf{l}\cdot\mathbf{b_{xx'}}/2} - e^{-i\mathbf{l}\cdot\mathbf{b_{xx'}}/2} \right] \left[e^{i\mathbf{l}\cdot\mathbf{b_{yy'}}/2} - e^{-i\mathbf{l}\cdot\mathbf{b_{yy'}}/2} \right]. \qquad (8.12)
$$

This is obtained from Eqs.(4.36) and (8.7) by substituting the defining relation Eq.(8.11) for each impact factor and integrating over $\mathbf{k_1}$ and $\mathbf{k_2}$ (these integrals just yield delta functions which fix the size of the parent dipole plus terms which vanish since $N(0, r, y, q) = 0$). The colour factor $\mathcal{G} = N^2 G_0^{(1)} = 2$.

Equation (8.12) has a very appealing physical interpretation. To see this let us first consider the amplitude in the approximation that the onia interact through the exchange of two gluons. In this case we have

$$
\frac{\Im m A(s,t)}{s} = \frac{2}{(2\pi)^4} \int \frac{d^2\mathbf{l}}{\mathbf{l}^2(\mathbf{l}-\mathbf{q})^2} \Phi_1(\mathbf{l},\mathbf{q})\Phi_2(\mathbf{l},\mathbf{q})
$$
$$
= \frac{8}{9}\alpha_s^2 \int dz_1 dz_2 \int d^2r_1 d^2r_2 |\Psi_1(z_1,r_1)|^2 |\Psi_2(z_2,r_2)|^2
$$
$$
\int \frac{d^2\mathbf{l}}{\mathbf{l}^2(\mathbf{l}-\mathbf{q})^2} \left[e^{i\mathbf{l}\cdot\mathbf{r_1}/2} - e^{-i\mathbf{l}\cdot\mathbf{r_1}/2} \right] \left[e^{i(\mathbf{q}-\mathbf{l})\cdot\mathbf{r_1}/2} - e^{-i(\mathbf{q}-\mathbf{l})\cdot\mathbf{r_1}/2} \right]
$$
$$
\left[e^{i\mathbf{l}\cdot\mathbf{r_2}/2} - e^{-i\mathbf{l}\cdot\mathbf{r_2}/2} \right] \left[e^{i(\mathbf{q}-\mathbf{l})\cdot\mathbf{r_2}/2} - e^{-i(\mathbf{q}-\mathbf{l})\cdot\mathbf{r_2}/2} \right]. \qquad (8.13)
$$

This equation is shown graphically in Fig. 8.1, where the factoriza-

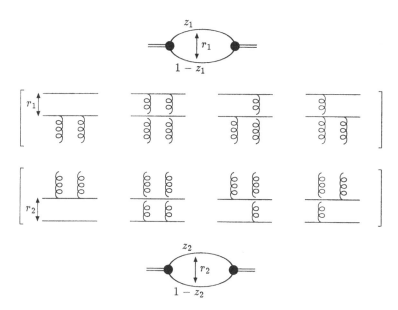

Fig. 8.1. Diagrammatic illustration of the dipole factorization explicit in Eq.(8.13).

tion of the onia wavefunctions from the dipole–dipole interaction cross-section is illustrated.

Comparing Eq.(8.13) with Eq.(8.12) we see some striking similarities. The only new factors are those associated with what we have termed the dipole number densities, N. In particular, the exponential factors are equivalent. This similarity means that we can interpret the elastic scattering of the two onia in terms of the scattering of individual dipoles in each onium state off those in the other state (since the dipole–dipole interaction cross-section is explicit in Eq.(8.12)). The number density of dipoles is then indeed given by the function $N(b, r, y, q)$. It specifies the number density of dipoles of size r inside an onium whose primary dipole (i.e. the q–$\bar{q}$ pair) has size b and which lie within y units of rapidity of the parent onium. The q-dependence is present since the number density depends upon the angle through which the onia scatter. Equivalently, we could take the Fourier transform with respect to q and obtain the number densities as a function of their displacement from the parent onia (indeed, we shall do this at the end of this section). The normalization has been chosen such that

$$\int \frac{d^2\mathbf{r}}{2\pi r^2} N(b, r, y, q)$$

is the total number of dipoles inside the parent dipole.

Note that the dipole number density at $q = 0$ and large enough y is the familiar looking expression (e.g. see Eq.(4.34)):

$$N(b, r, y, 0) \approx \frac{b}{r} \frac{e^{\omega_0 y}}{(\pi a^2 y)^{1/2}} \exp\left(-\frac{\ln^2(b^2/r^2)}{4a^2 y}\right). \qquad (8.14)$$

In addition, we note that it is not meaningful to associate in a unique way the dipoles with the colliding onia. By picking $y' = 0$, $N(r_1, b_{xx'}, 0, q) = r_1\delta(r_1 - b_{xx'})$ and all the dipoles are identified as being radiated from the parent dipole of size r_2 (i.e. from the onium which has wavefunction Ψ_2). By picking $y' = y/2$ we have the democratic scenario where the dipoles are 'shared' between the colliding onia.

So, we have been forced into the interpretation of onium–onium scattering in terms of the interaction between 'child' dipoles (the parents being the q–$\bar{q}$ pairs of the onia); the interaction being none other than the two-gluon exchange between the two child dipoles. It is natural to ask how these dipoles originate. We know from the above that we have succeeded in factorizing the BFKL physics associated with the ladder of reggeized gluons into the dipole density functions. What has happened? Let us think in a frame where the two onia are colliding in their centre of mass. We can then identify a left-moving onium and a right-moving onium. Now consider the left-moving onium. The parent dipole is created a long time before the interaction with a dipole in the other onium. This parent dipole can then radiate a soft gluon. In terms of its colour structure the emitted gluon can be viewed as a $3 \odot \bar{3}$ state (i.e. like a quark–antiquark pair). This view of the gluon is appropriate in the formal limit where the number of colours, N, is large. This means that the leading logarithm approximation is also the leading N approximation – indeed we can see this by noting that the relevant coupling for soft gluon emission is $\bar{\alpha}_s$ (i.e. the strong coupling is always accompanied by a factor of N). The quark line from this gluon and the antiquark line of the parent dipole then form a secondary dipole, and similarly for the quark line of the gluon and the antiquark line of the parent. *So the emission of a gluon corresponds to the annihilation of the parent dipole and the creation of two new secondary dipoles.* This branching continues until there is no more rapidity for further emission and so the parent left-moving onium can be viewed as a collection of colour dipoles. The same can be said about the right-moving onium: it, too, is an assembly of dipoles. The onia then interact with each other through the scattering of their constituent dipoles which, in the one Pomeron exchange approximation (which is the BFKL approximation), scatter via exchange of two gluons. The left and right-moving dipoles then re-assemble to generate the final state onia. In this way we are able to understand the origin of the dipole factorization which is explicit in Eq.(8.12).

Before leaving this section, let us first re-cast Eq.(8.12) so that it explicitly exhibits the dependence of the scattering amplitude on the impact parameter of the collision (i.e. we take a Fourier transform to eliminate the **q**-dependence).

Defining the scattering amplitude for collisions between two onia at impact parameter **b** via

$$A(b, y) = \int \frac{d^2\mathbf{q}}{(2\pi)^2} e^{-iq\cdot b} \frac{A(s, t)}{2s} \tag{8.15}$$

allows us to write

$$
\begin{aligned}
A(b, y) &= -i\frac{8}{9} \int dz_1 dz_2 \int d^2\mathbf{r_1} d^2\mathbf{r_2} |\Psi_1(z_1, r_1)|^2 |\Psi_2(z_2, r_2)|^2 \\
&\times F(r_1, r_2, b, y),
\end{aligned}
\tag{8.16}
$$

where $F(r_1, r_2, b, y)$ is the amplitude for the elastic scattering of a pair of dipoles of respective sizes r_1 and r_2 at an impact parameter b. Explicitly it is given by

$$
\begin{aligned}
F(r_1, r_2, b, y) &= -\int \frac{d^2\mathbf{b_{xx'}}}{2\pi \mathbf{b_{xx'}}^2} \int \frac{d^2\mathbf{b_{yy'}}}{2\pi \mathbf{b_{yy'}}^2} \int d^2\mathbf{R} d^2\mathbf{R'} \\
&\times n(r_1, b_{xx'}, y', R) n(r_2, b_{yy'}, y - y', |\mathbf{R'} - \mathbf{b}|) \\
&\times f(\mathbf{R} - \mathbf{R'}, \mathbf{b_{xx'}}, \mathbf{b_{yy'}}),
\end{aligned}
\tag{8.17}
$$

where

$$
\begin{aligned}
f(\mathbf{R}, \mathbf{b}, \mathbf{c}) \;=\;& \frac{\alpha_s^2}{2} \int \frac{d^2\mathbf{q}}{(2\pi)^2} e^{-i\mathbf{q}\cdot\mathbf{R}} \int \frac{d^2\mathbf{l}}{\mathbf{l}^2(\mathbf{l}-\mathbf{q})^2} \\
&\times \left[e^{i(\mathbf{q}-\mathbf{l})\cdot\mathbf{b}/2} - e^{-i(\mathbf{q}-\mathbf{l})\cdot\mathbf{b}/2} \right] \left[e^{i(\mathbf{q}-\mathbf{l})\cdot\mathbf{c}/2} - e^{-i(\mathbf{q}-\mathbf{l})\cdot\mathbf{c}/2} \right] \\
&\times \left[e^{i\mathbf{l}\cdot\mathbf{b}/2} - e^{-i\mathbf{l}\cdot\mathbf{b}/2} \right] \left[e^{i\mathbf{l}\cdot\mathbf{c}/2} - e^{-i\mathbf{l}\cdot\mathbf{c}/2} \right].
\end{aligned}
\tag{8.18}
$$

The number density of dipoles, of size $\mathbf{x}$ within a parent dipole of size $\mathbf{x_0}$ within the rapidity y and at a distance $\mathbf{r}$ of the parent is $n(x_0, x, y, r)$, where

$$
n(x_0, x, y, r) = \int \frac{d^2\mathbf{q}}{(2\pi)^2} e^{-i\mathbf{q}\cdot\mathbf{r}} N(x_0, x, y, q).
\tag{8.19}
$$

Representing the amplitude exclusively in terms of the positions and sizes of the dipoles will be convenient when we come to discuss the multiple scattering corrections in the next section. For now, let us express the optical theorem in terms of $A(b, y)$; it is simply

$$
\sigma_{\text{tot}}(y) = 2\pi \int db^2 \, \Im m A(b, y).
\tag{8.20}
$$

Our normalization of the amplitude is such that $\Im m A(b, y) = \theta(b_0 - b)$ in the black disc limit (i.e. totally absorbtive scattering).

To close this section, we note that by taking Eqs.(8.8) and (8.9), expanding $\chi_0(\nu)$ up to quadratic order in ν and integrating over ν using the saddle point approximation we arrive at the approximation

$$
\begin{aligned}
n(x_0, x, y, r) \;\approx\;& \frac{2x_0}{xr^2} \ln\left(\frac{16r^2}{xx_0} \right) \frac{e^{\omega_0 y}}{(\pi a^2 y)^{3/2}} \\
&\times \exp\left(-\frac{\ln^2(16r^2/xx_0)}{a^2 y} \right),
\end{aligned}
\tag{8.21}
$$

provided $r \gg x, x_0$ and $a^2 y \gg \ln(r^2/xx_0)$. Compare this with the result of the preceding chapter, Eq.(7.14). The diffusion properties of the BFKL equation are manifest as diffusion in the dipole sizes with increasing rapidity. The displacement of the child dipole from the parent acts as an effective cut-off on the size of the largest dipoles that can be created.

8.2 Unitarity

One of the main results which is obtained in the leading logarithmic approximation used to derive the BFKL equation is that the elastic scattering amplitudes rise with increasing centre-of-mass energy, s, as some power of s. Through the optical theorem this then translates into a corresponding growth of the total cross-section, i.e.

$$\sigma_{\text{tot}} \sim s^{\omega_0}, \tag{8.22}$$

where $\omega_0 = 4\bar{\alpha}_s \ln 2$. We should ask whether this is sensible behaviour in the limit $s \to \infty$. Intuitively, if the strong interactions are of finite range then we expect the asymptotic behaviour of total cross-sections to be limited in some way. This physics is missing in the leading logarithmic approximation. Moreover, in Chapter 1 we quoted the Froissart–Martin bound, which states that total cross-sections cannot rise (in the limit $s \to \infty$) faster than $\ln^2 s$ (see Eq.(1.25)). Although this bound may well become significant only at energies well beyond those which are feasibly accessible it is important to understand how the leading logarithmic approximation is corrected to account for the unitarization corrections which eventually bring the theory into agreement with the Froissart–Martin bound. The study of unitarity corrections within perturbative QCD is the subject of the remainder of this chapter.

We start by providing a physical argument (originally due to Feynman) which makes the Froissart–Martin bound plausible. Let us suppose that the target particle has some density distribution which reflects the short range nature of the strong force, e.g.

$$\rho(r) = \rho_0 \exp(-r/R), \tag{8.23}$$

where r is the distance from the centre of the target and R characterizes the size of the target. It is important that this distribution falls off faster than any power at large distances (which we take as a fundamental property of the strong interactions). If the probability of an interaction between the beam particle with the target is bounded (as $s \to \infty$) by some finite power of s then the interaction probability satisfies

$$P(s, r) < P_0 \left(\frac{s}{s_0}\right)^N \exp(-r/R). \tag{8.24}$$

Hence, the interaction will be negligible for collisions at impact parameters

$$r > NR \ln (s/s_0)$$

and so the total cross-section satisfies

$$\sigma_{\text{tot}} < \pi R^2 N^2 \ln^2(s/s_0). \qquad (8.25)$$

It is possible to derive the Froissart–Martin bound in a more rigorous fashion starting from the partial wave expansion and assuming the amplitude to satisfy (subtracted) dispersion relations (this is the assumption that the amplitude is bounded by a finite power in s). It arises as a direct consequence of the unitarity of the individual partial wave amplitudes and the existence of some lowest mass bound state whose mass is different from zero (i.e. the pion) (see e.g. Collins (1977), Martin, Morgan & Shaw (1976)). This latter property, which is equivalent to demanding that the strong force be short range, is one which we do not expect to be able to accommodate in our perturbative calculations, as such we might well be able to successfully unitarize the scattering amplitude but fail to satisfy the Froissart–Martin bound.

Clearly, therefore, all our previous calculations based on QCD in the leading logarithm approximation must break down as the centre-of-mass energy tends to infinity. In the centre-of-mass frame of the colliding particles the increase of the total cross-section with energy is due to the proliferation of soft gluon emissions. The power-like increase in the number of soft gluons means a corresponding rise in the total cross-section. In the dipole language it is the proliferation of dipoles which drives the rise. It is not hard to imagine the physics which must eventually enter as the spatial density of gluons (dipoles) continues to increase. Ultimately, the density will be large enough such that *more than one pair of dipoles will undergo a scattering* for each parent particle collision. There is also the possibility that a dipole in the parent can scatter off other dipoles also within the parent. As we shall see the distinction between these two forms of correction is frame dependent. Not surprisingly both forms of correction lead to a taming of the growth of the elastic scattering amplitude (and hence total cross-section) in line with the demands of unitarity. It is the purpose of the remainder of this chapter to describe the multiple scattering mechanism in more detail.

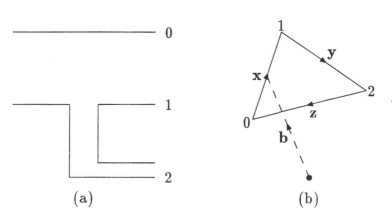

Fig. 8.2. (a) Fundamental dipole vertex used to generate the dipole evolution. A gluon is emitted at position 2. (b) The vectors which specify the size and position of the parent and child dipoles.

8.2.1 The operator formalism

We start by introducing an operator formalism (Mueller (1995)) which can be used to re-derive the BFKL equation but which will also be suitable for a quantitative study of the leading multiple scattering corrections. We have shown that high energy scattering between two onia can be viewed as a two-step process. Firstly, a cloud of dipoles is evolved around each of the primary dipoles. This dipole evolution can be described as a classical branching process in impact parameter space. Secondly, the dipole clouds interact with each other so that the total cross-section is an incoherent sum over the individual dipole–dipole cross-sections. The nature of the dipole branching process suggests that we should be able to describe it using an operator formalism where the basic operators are dipole creation and annihilation operators. There is a fundamental vertex which describes the branching of an initial dipole into two secondary dipoles and it is the successive iteration of this basic vertex that determines the evolution of the dipole cloud.

We begin by deriving an expression for the fundamental dipole vertex illustrated in Fig. 8.2(a). The parent dipole (specified by the points denoted 0 and 1) radiates a gluon at point 2 which generates the child dipoles (of sizes y and z, see Fig. 8.2(b)). Firstly, we introduce dipole creation and annihilation operators $a^\dagger(\mathbf{b}, \mathbf{x})$ and $a(\mathbf{b}, \mathbf{x})$ respectively ($\mathbf{b}$ is the location of the dipole centre and $\mathbf{x}$ is its size). Since the dipoles satisfy bosonic statistics we impose the commutation relation

$$[a(\mathbf{b}, \mathbf{x}), a^\dagger(\mathbf{b}', \mathbf{x}')] = \delta^{(2)}(\mathbf{b} - \mathbf{b}')\delta^{(2)}(\mathbf{x} - \mathbf{x}'). \qquad (8.26)$$

The differential probability for the emission of a gluon off a dipole of size $\mathbf{r}$ into the rapidity interval $y \to y + dy$ and with transverse momentum $\mathbf{k} \to \mathbf{k} + d^2\mathbf{k}$ is

$$d^3 P = \bar{\alpha}_s \frac{d^2\mathbf{k}}{\pi \mathbf{k}^2} dy(1 - e^{i\mathbf{k}\cdot\mathbf{r}}). \qquad (8.27)$$

For the derivation of this expression we refer to Appendix B of this chapter. The colour factor is appropriate for purely gluonic branching, i.e. we need an adjustment which accounts for the different coupling to the primary dipole, which is a q-$\bar{q}$ pair. This is the origin of the colour factor of $\frac{8}{9}$ which sits outside the amplitude, e.g. Eq.(8.16).[†] This very simple form is, however, unsuitable for the dipole evolution. We need to obtain an expression in terms of the relevant dipole sizes. Starting from

$$\int \frac{d^2\mathbf{k}}{\mathbf{k}^2} e^{i\mathbf{k}\cdot\mathbf{r}} = \int d^2\mathbf{k} \frac{\mathbf{k} \cdot \mathbf{k}}{\mathbf{k}^4} e^{i\mathbf{k}\cdot\mathbf{r}} \qquad (8.28)$$

and using

$$\frac{\mathbf{k}_i}{\mathbf{k}^2} = \frac{1}{2\pi i} \int \frac{d^2\mathbf{x}}{\mathbf{x}^2} \mathbf{x}_i e^{i\mathbf{k}\cdot\mathbf{x}} \qquad (8.29)$$

yields

$$\int \frac{d^2\mathbf{k}}{\mathbf{k}^2} e^{i\mathbf{k}\cdot\mathbf{r}} = -\int d^2\mathbf{x}_1 d^2\mathbf{x}_2 \frac{\mathbf{x}_1 \cdot \mathbf{x}_2}{\mathbf{x}_1^2 \mathbf{x}_2^2} \delta^{(2)}(\mathbf{x}_1 + \mathbf{x}_2 + \mathbf{r}). \qquad (8.30)$$

Thus we can write

$$\begin{aligned}
\frac{d^3 P}{dy} &= \frac{\bar{\alpha}_s}{2\pi} d^2\mathbf{x}_1 d^2\mathbf{x}_2 \delta^{(2)}(\mathbf{x}_1 + \mathbf{x}_2 + \mathbf{r}) \left(\frac{(\mathbf{x}_1 + \mathbf{x}_2)^2}{\mathbf{x}_1^2 \mathbf{x}_2^2}\right) \\
&= \frac{\bar{\alpha}_s}{2\pi} d^2\mathbf{x}_1 d^2\mathbf{x}_2 \delta^{(2)}(\mathbf{x}_1 + \mathbf{x}_2 + \mathbf{r}) \left(\frac{r^2}{x_1^2 x_2^2}\right). \qquad (8.31)
\end{aligned}$$

[†] This factor is equal to $1 - 1/N^2$ and as such is equal to unity in the leading N approximation where there is no difference between the colour structure of the gluon and that of a q-$\bar{q}$ pair.

This form is now suitable for use in the dipole evolution since it describes the branching of the parent dipole of size r into two new dipoles of sizes x_1 and x_2. Each of these secondary dipoles can then act as a source for further gluon emission, the probability of which can be computed using Eq.(8.31), and so on.

In the operator language, the basic vertex for dipole creation is thus

$$V_1[a^\dagger, a] = \frac{\bar{\alpha}_s}{2\pi} \int d^2\mathbf{b}\, d^2\mathbf{x}\, d^2\mathbf{y}\, d^2\mathbf{z}\, \delta^{(2)}(\mathbf{x} + \mathbf{y} + \mathbf{z})$$

$$\times \frac{x^2}{y^2 z^2} a^\dagger(\mathbf{b} + \mathbf{y}/2, \mathbf{z}) a^\dagger(\mathbf{b} - \mathbf{z}/2, \mathbf{y}) a(\mathbf{b}, \mathbf{x}). \tag{8.32}$$

The square brackets indicate that it is a functional of the creation and annihilation operators. The dipole evolution is driven by this vertex. The arguments of the dipole operators can be seen from Fig. 8.2(b). However, things are not quite so simple. Recall that the BFKL equation contains essential virtual corrections. These corrections are so far absent. However, we can construct the correction, V_2, to the basic vertex, V_1, which accounts for all the virtual graphs. The vertex V_1 possesses ultra-violet divergences whenever the emitted dipoles have vanishing size (i.e. $y \to 0$ or $z \to 0$). In order to regularize these divergences we introduce a lower cut-off ρ on the size of the emitted dipoles. The virtual graphs are accounted for through the vertex

$$V_2[a^\dagger, a] = -\frac{\bar{\alpha}_s}{2\pi} \int d^2\mathbf{b}\, d^2\mathbf{x}\, d^2\mathbf{y}\, d^2\mathbf{z}\, \delta^{(2)}(\mathbf{x} + \mathbf{y} + \mathbf{z})$$

$$\times \frac{x^2}{y^2 z^2} a^\dagger(\mathbf{b}, \mathbf{x}) a(\mathbf{b}, \mathbf{x})$$

$$\approx -\bar{\alpha}_s \int d^2\mathbf{b}\, d^2\mathbf{x} \ln \frac{x^2}{\rho^2} a^\dagger(\mathbf{b}, \mathbf{x}) a(\mathbf{b}, \mathbf{x}). \tag{8.33}$$

This form is determined by requiring the conservation of probability, i.e. the total probability to create a pair of secondary dipoles integrated over all the sizes of these dipoles, plus the probability *not* to create a secondary pair must be unity. We demonstrate this to order α_s below. The approximate equality on the second line of Eq.(8.33) is valid in the limit of small ρ. The complete vertex for dipole evolution is then

$$V[a^\dagger, a] = V_1[a^\dagger, a] + V_2[a^\dagger, a]$$

and is independent of ρ (as $\rho \to 0$).

We are now in a position to construct the S-matrix for the scattering of primary dipoles of sizes r_1 and r_2 (it will give us $F(r_1, r_2, b, y)$ of Eq.(8.17)). Consider the centre-of-mass scattering of the two primary dipoles (we refer to them as left-moving and right-moving). The left-moving dipole (at position $\mathbf{b_0}$ and of size r_1) is the state

$$|\mathbf{b_0}, r_1\rangle = a^\dagger(\mathbf{b_0}, r_1)|0\rangle, \qquad (8.34)$$

where $\langle 0|0\rangle = 1$. Similarly, the right-moving dipole (which is at impact parameter $\mathbf{b}$ relative to the left mover) is the state

$$|\mathbf{b} + \mathbf{b_0}, r_2\rangle = d^\dagger(\mathbf{b} + \mathbf{b_0}, r_2)|0\rangle, \qquad (8.35)$$

where d and $d^\dagger$ are the annihilation and creation operators for the right movers (we need independent operators since the two dipole clouds evolve independently, i.e. the left mover operators commute with the right mover operators). The probability of finding the primary left mover dipole in a configuration of n dipoles with positions and sizes $\{\mathbf{b_1}, \mathbf{c_1}; \mathbf{b_2}, \mathbf{c_2}; \cdots; \mathbf{b_n}, \mathbf{c_n}\}$ is thus

$$\frac{d^{4n}p_n}{d^2\mathbf{b_1} d^2\mathbf{c_1} \cdots d^2\mathbf{b_n} d^2\mathbf{c_n}}$$
$$= \langle 0|a(\mathbf{b_1}, \mathbf{c_1}) \cdots a(\mathbf{b_n}, \mathbf{c_n}) e^{yV_L} a^\dagger(\mathbf{b_0}, r_1)|0\rangle. \qquad (8.36)$$

We have used the subscript L to denote that the vertex operator acts on left movers. The basic vertex appears in the exponential due to the combinatorial factorial factor which is needed on iterating the basic vertex (recall that the vertex integrates over all dipole configurations). A similar expression exists for the evolution of the right movers. It is convenient to define the n-dipole state (integrated over all dipole locations and sizes):

$$|n\rangle \equiv \frac{1}{n!} \int \left(\prod_{j=1}^{n} d^2\mathbf{c_j} d^2\mathbf{b_j} a^\dagger(\mathbf{b_j}, \mathbf{c_j}) \right) |0\rangle. \qquad (8.37)$$

The integrated probability for the n-dipole configuration is then

$$P_n = \frac{1}{n!} \int d^{4n}p_n$$
$$= \langle n|e^{yV_L} a^\dagger(\mathbf{b_0}, r_1)|0\rangle, \qquad (8.38)$$

and satisfies

$$\sum_{n=1}^{\infty} P_n = 1, \tag{8.39}$$

which is consistent with our interpretation of P_n as a probability. Working to first order in α_s allows us to see the conservation of probability and the role of the virtual corrections (V_2) explicitly. At this order, only P_1 and P_2 are non-zero and it is easy to show that

$$P_1 = 1 - \bar{\alpha}_s \ln \frac{r_1^2}{\rho^2},$$

$$P_2 = \bar{\alpha}_s \ln \frac{r_1^2}{\rho^2}. \tag{8.40}$$

The dependence upon the ultra-violet cut-off (ρ) cancels, as required. Moreover, the virtual corrections generated by V_2 are solely responsible for the logarithmic term in P_1 which ensures the conservation of probability.

The scattering matrix for the elastic scattering of the left- and right-moving primary dipoles is given by

$$S(\mathbf{r_1}, \mathbf{r_2}, \mathbf{b}, y) =$$
$$\langle 0 | e^{a_1 + d_1} e^{-f} e^{y'V_L + (y - y')V_R} d^\dagger(\mathbf{b} + \mathbf{b_0}, \mathbf{r_2}) a^\dagger(\mathbf{b_0}, \mathbf{r_1}) | 0 \rangle, \tag{8.41}$$

where

$$a_1 = \int d^2 \mathbf{b} d^2 c \, a(\mathbf{b}, \mathbf{c}) \tag{8.42}$$

(and similarly for d_1). The dipole–dipole scattering operator, f, is given by

$$f = \int d^2 \mathbf{R} d^2 \mathbf{R}' d^2 c d^2 c' f(\mathbf{R} - \mathbf{R}', \mathbf{c}, \mathbf{c}')$$
$$\times d^\dagger(\mathbf{R}, \mathbf{c}) d(\mathbf{R}, \mathbf{c}) a^\dagger(\mathbf{R}', \mathbf{c}') a(\mathbf{R}', \mathbf{c}') \tag{8.43}$$

and $f(\mathbf{R} - \mathbf{R}', \mathbf{c}, \mathbf{c}')$ is given in Eq.(8.18). A few words are in order regarding Eq.(8.41). Starting from the 'vacuum state' on the right, we first create the primary dipoles (of size $\mathbf{r_1}$ and $\mathbf{r_2}$ with relative separation $\mathbf{b}$). The action of the dipole evolution operators then generates the respective dipole clouds. These clouds are then made to interact. The amplitude for any single dipole–dipole interaction is given by $-f$ (the operator structure of Eq.(8.43) is such that it projects out the dipoles that interact from the evolved dipole clouds). If we assume that there are very many dipoles and that these dipoles scatter independently of each other then we are able to account for an arbitrary number of individual dipole–dipole interactions by including the factor $(-f)^n/n!$ for n dipole–dipole interactions. This explains the origin of the factor e^{-f} in Eq.(8.41). The final factor simply ensures that the dipole systems have unit overlap with the final state. The factorial factors associated with the various exponential terms are necessary in order to divide out the equivalent configurations (recall that all secondary dipole configurations are integrated over to determine the elastic amplitude). Thus we have a formalism which allows us to include the **multiple scattering** of individual dipoles. As we shall soon see, the BFKL (leading logarithmic approximation) is equivalent to including only the interaction of a single pair of dipoles (which eventually violates unitarity) whilst the complete multiple scattering series ensures that unitarity is preserved.

We can re-write the S-matrix in an alternative form by inserting the unit operator,

$$\sum_{m,n} |n,m\rangle\langle n,m|, \qquad (8.44)$$

where the state $|n,m\rangle = |n\rangle|m\rangle$ represents n dipoles in the left-moving onium and m dipoles in the right-moving onium. We find

$$S(\mathbf{r_1},\mathbf{r_2},\mathbf{b},y) = \sum_{m,n} P_n P_m \exp(-\langle n,m|f|n,m\rangle). \qquad (8.45)$$

Note that this expression explicitly satisfies the constraints of unitarity. To see this we note that $|1 - S|^2$ is the probability of an elastic scattering occurring at a fixed impact parameter and as such should satisfy $|1 - S|^2 \leq 1$. This bound is indeed satisfied since the P_n and P_m are probabilities and because $\langle n,m|f|n,m\rangle$ is positive definite (see Eq.(8.43)). This is not the case for single Pomeron exchange, where the e^{-f} factor is replaced by $1 - f$.

In the one-Pomeron exchange approximation, the formalism we have just described must be completely equivalent to the BFKL (leading $\ln s$) one, i.e. replacing the e^{-f} factor by $1 - f$ should lead to

$$S_1(\mathbf{r_1},\mathbf{r_2},\mathbf{b},y) = 1 + F(r_1,r_2,b,y), \qquad (8.46)$$

where $F(r_1,r_2,b,y)$ is defined in Eq.(8.17). It is enlightening to spend a little time outlining the proof of this equivalence.

We start by quoting the result (it is not hard to show):
$$e^{a_1} f[a^\dagger] e^{-a_1} = f[a^\dagger + 1],\qquad(8.47)$$
where $f[a^\dagger]$ is some functional of the creation operator. Using Eq.(8.47), we can re-write Eq.(8.41) in the one-Pomeron exchange approximation as

$$\begin{aligned}
S_1(r_1, r_2, b, y) &= 1 - \int d^2\mathbf{R}\, d^2\mathbf{R}'\, d^2\mathbf{c}\, d^2\mathbf{c}'\, f(\mathbf{R} - \mathbf{R}', \mathbf{c}, \mathbf{c}') \\
&\times \langle 0| a(\mathbf{R}', \mathbf{c}') e^{y' V_L[a^\dagger+1, a]} a^\dagger(\mathbf{b_0}, \mathbf{r_1})|0\rangle \\
&\times \langle 0| d(\mathbf{R}, \mathbf{c}) e^{(y-y') V_R[d^\dagger+1, d]} d^\dagger(\mathbf{b} - \mathbf{b_0}, \mathbf{r_2})|0\rangle.
\end{aligned}\qquad(8.48)$$

which, on comparison with Eq.(8.17), reveals that
$$\frac{n(r_1, c', y', R')}{2\pi c'^2} = \langle 0| a(\mathbf{R}', \mathbf{c}') e^{y' V_L[a^\dagger+1, a]} a^\dagger(\mathbf{b_0}, \mathbf{r_1})|0\rangle.\qquad(8.49)$$
We need to evaluate explicitly the right hand side of this expression and demonstrate its equivalence to the number density calculated using Eqs.(8.8) and (8.19).

Since the only terms in the exponent which generate a non-zero contribution to the number density are those which contain equal numbers of creation and annihilation operators we can make the replacement
$$V[1 + a^\dagger, a] \to a^\dagger K a.\qquad(8.50)$$
Selecting terms in $V_1[1+a^\dagger, a]+V_2[a^\dagger, a]$ (see Eqs.(8.32) and (8.33)) that contain one creation and one annihilation operator we find
$$a^\dagger K a \equiv \int d^2\mathbf{b}\, d^2\mathbf{x}\, d^2\mathbf{x}'\, d^2\mathbf{b}'\, a^\dagger(\mathbf{b} + \mathbf{b}', \mathbf{x}) K(\mathbf{b}', \mathbf{x}, \mathbf{x}') a(\mathbf{b}, \mathbf{x}')$$
$$(8.51)$$
and the evolution kernel is (using Eqs.(8.32) and (8.33))

$$\begin{aligned}
K(\mathbf{b}', \mathbf{x}, \mathbf{x}') &= -\bar{\alpha}_s \ln\frac{x^2}{\rho^2} \delta^{(2)}(\mathbf{b}') \delta^{(2)}(\mathbf{x} - \mathbf{x}') \\
&+ \frac{\bar{\alpha}_s}{8\pi} \frac{x'^2}{b'^2 x^2} \left[\delta^{(2)}(\mathbf{b}' + (\mathbf{x} + \mathbf{x}')/2) + \delta^{(2)}(\mathbf{b}' - (\mathbf{x} + \mathbf{x}')/2) \right].
\end{aligned}\qquad(8.52)$$

The eigenfunctions of this operator are none other than the conformal eigenfunctions, i.e.

$$\int d^2\mathbf{b}\, \frac{d^2\mathbf{x}'}{\mathbf{x}'^4} K(\mathbf{b}' - \mathbf{b}, \mathbf{x}, \mathbf{x}') \tilde{\phi}_n^\nu(\mathbf{b} + \mathbf{x}'/2, \mathbf{b} - \mathbf{x}'/2, \mathbf{w})$$
$$= \bar{\alpha}_s \chi_n(\nu) \frac{\tilde{\phi}_n^\nu(\mathbf{b}' + \mathbf{x}/2, \mathbf{b}' - \mathbf{x}/2, \mathbf{w})}{\mathbf{x}^4}\qquad(8.53)$$

and the eigenvalues are the eigenvalues of the BFKL kernel. To derive this important result, it is best to move to complex co-ordinates (i.e. $2d^2\mathbf{x} = dx\,d\bar{x}$) whereupon the integrand separates into a product of one-dimensional integrals over x and $\bar{x}$. After a change of variables the integrals can be rewritten in two-vector form, where they are seen to generate the eigenfunctions using the result that[†]

$$\int \frac{d^2\mathbf{z}}{2\pi} \frac{[2(\mathbf{z}^2)^{1/2+i\nu} - 1]}{\mathbf{z}^2(\mathbf{z} + \hat{\mathbf{n}})^2} = \chi_0(\nu), \qquad (8.54)$$

where $\hat{\mathbf{n}}$ is an arbitrary unit vector. This result suggests that we should expand the dipole creation and annihilation operators in terms of these eigenfunctions, i.e.

$$a(\mathbf{b}, \mathbf{x}) = \sum_{n=-\infty}^{\infty} \int \frac{d\nu}{(2\pi)^2} 4(i\nu + n/2) a_{n\nu}(\mathbf{w}) \frac{d^2\mathbf{w}}{\mathbf{x}^4}$$
$$\times \tilde{\phi}_n^{\nu}(\mathbf{b} + \mathbf{x}/2, \mathbf{b} - \mathbf{x}/2, \mathbf{w}), \qquad (8.55)$$

and

$$a^{\dagger}(\mathbf{b}, \mathbf{x}) = \sum_{n=-\infty}^{\infty} \int \frac{d\nu}{(2\pi)^2} 4(-i\nu + n/2) a_{n\nu}^{\dagger}(\mathbf{w}) d^2\mathbf{w}$$
$$\times \tilde{\phi}_n^{\nu*}(\mathbf{b} + \mathbf{x}/2, \mathbf{b} - \mathbf{x}/2, \mathbf{w}). \qquad (8.56)$$

Using Eq.(4.50) the 'conformal' operators can be shown to satisfy the commutation relation

$$[a_{n\nu}(\mathbf{w}), a_{n'\nu'}^{\dagger}(\mathbf{w}')] = \delta_{nn'}\delta(\nu - \nu')\delta^{(2)}(\mathbf{w} - \mathbf{w}') \qquad (8.57)$$

and, using the known properties of the eigenfunctions (Eqs.(8.4) and (8.5)), we can recast the evolution operator in the diagonal form

$$a^{\dagger}Ka = \bar{\alpha}_s \sum_{n=-\infty}^{\infty} \int d\nu\, d^2\mathbf{w}\, \chi_n(\nu) a_{n\nu}^{\dagger}(\mathbf{w}) a_{n\nu}(\mathbf{w}). \qquad (8.58)$$

Using Eqs.(8.55), (8.56), (8.57) and (8.58) in Eq.(8.49) allows us to show that (in the $n = 0$ case)

$$n(r_1, c, y, R) = 16 \int \frac{d\nu}{(2\pi)^3} \frac{d^2\mathbf{w}}{c^2} \nu^2 e^{\bar{\alpha}_s \chi_0(\nu)y}$$
$$\tilde{\phi}_0^{\nu}(\mathbf{R} + \mathbf{c}/2, \mathbf{R} - \mathbf{c}/2, \mathbf{w}) \tilde{\phi}_0^{\nu*}(\mathbf{b_0} + \mathbf{r_1}/2, \mathbf{b_0} - \mathbf{r_1}/2, \mathbf{w}). \quad (8.59)$$

[†] This is appropriate for $n = 0$ but it is not much more difficult to prove the result for general n.

This is equivalent to the result that is obtained in a straightforward way after substituting Eq.(8.8) into Eq.(8.19).

Thus we have demonstrated the equivalence of the operator formalism to that of BFKL. Indeed, the dipole formalism does offer an alternative derivation of the BFKL equation (Mueller (1994, 1995), Mueller & Patel (1994), Chen & Mueller (1995), Nikolaev, Zakharov & Zoller (1994a,b), Nikolaev & Zakharov (1994)). However, more than merely reproducing the results obtained in Chapter 4 we have now established a framework in which we can investigate the multiple scattering corrections which motivated this alternative approach, i.e. we can go beyond the one-Pomeron exchange approximation.

8.2.2 Multiple scattering

Consider the total cross-section for the scattering of two primary dipoles of fixed (and equal) sizes R (this avoids us having to invoke specific onium wavefunctions and should demonstrate all the important features). It is natural to ask when the one Pomeron exchange approximation (BFKL) starts to break down. Formally the S-matrix for the elastic scattering can be written as a multiple scattering series; keeping only the first term corresponds to the BFKL calculation and has the S-matrix of Eq.(8.46). This will only be a good approximation provided $|F| \ll 1$. When $|F| \sim 1$ it becomes necessary to consider the remaining terms in the multiple scattering series. We shall discuss these terms shortly, but for now we have a simple condition for the validity of the BFKL calculation. We can evaluate $F(R, R, b, y)$ in the limit of $b^2 \gg R^2$ and for $a^2 y \gg \ln(b^2/R^2)$ (by the usual saddle point method). Our condition for the legitimate neglect of the multiple scattering corrections then becomes the explicit condition

$$-F(R, R, b, y) = 8\pi\alpha_s^2 \frac{R^2}{b^2} \frac{\ln(b^2/R^2)}{(\pi a^2 y)^{3/2}} e^{\omega_0 y} \exp\left(-\frac{\ln^2(b^2/R^2)}{a^2 y}\right)$$

$$\ll 1. \tag{8.60}$$

The total cross-section is formed using Eq.(8.20), i.e.

$$\sigma_{\text{tot}} = -2\pi \int db^2 F(R, R, b, y) = 8\pi R^2 \frac{\alpha_s^2 e^{\omega_0 y}}{(\pi a^2 y)^{1/2}}. \tag{8.61}$$

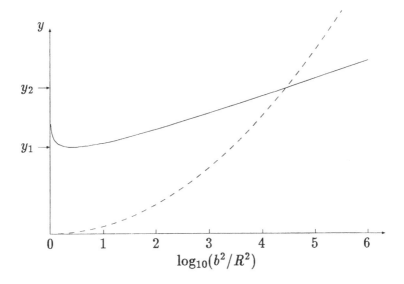

Fig. 8.3. Plot to delineate the region where the one-Pomeron exchange (BFKL) calculation (of the total cross-section) is valid from the region where multiple scattering is important.

The dominant contribution to this cross-section comes from the region of large b, in particular

$$b^2 \sim R^2 e^{\sqrt{a^2 y}} \qquad (8.62)$$

and we have a self-consistent calculation (i.e. Eq.(8.60) is valid in the region which gives rise to the dominant contribution to the total cross-section). Note that the total cross-section is driven by the contribution from peripheral collisions (i.e. $b \gg R$). The inequality of Eq.(8.60) can be re-written as a bound on y at a given impact parameter, i.e. defining $y(b)$ to be the solution to $-F(R, R, b, y) = 1$ we find

$$\omega_0 y(b) \approx \ln \left(\frac{(\pi a^2 y(b))^{3/2}}{8\pi \alpha_s^2} \frac{b^2 / R^2}{\ln(b^2 / R^2)} \right) \qquad (8.63)$$

and the condition for the validity of the one Pomeron exchange calculation of the amplitude at some impact parameter b is then that $y \ll y(b)$.

In Fig. 8.3, the solid line corresponds to the curve $y = y(b)$, whilst the dashed line corresponds to $y = (1/a^2)\ln^2(b^2/R^2)$ (which, from Eq.(8.62), specifies the region which provides the dominant contribution to the total cross-section). We define y_1 to correspond to the minimum of the curve $y = y(b)$ and y_2 to be the rapidity where the two curves intersect. For $y < y_1$ the one Pomeron exchange approximation is appropriate over the whole range of impact parameter which contributes to the total cross-section and so we can trust the BFKL calculation in this region. For $y_1 < y < y_2$ multiple scattering corrections are significant for a wide range of impact parameter. However, the dominant contribution to the total cross-section still arises from the region of large impact parameter where the BFKL calculation is again valid. For $y > y_2$ multiple scattering corrections are now significant even in the region which contributes most to the total cross-section. Thus only for $y > y_2$ do we need to worry about the role of multiple scattering (unitarization) corrections to the total cross-section. Fig. 8.3 was produced with $\bar{\alpha}_s = 0.25$, in which case $y_1 \approx 15$, which is quite large and indicates that unitarity corrections to the total cross-section are important only at very high energies. The slow onset of the unitarity corrections is due essentially to the peripheral nature of the dominant contributions to the total rate, i.e. multiple scattering effects are most important for the more central collisions (where there is a large overlap between the left and right-moving dipole clusters).

A process which is more sensitive to the multiple scattering corrections will therefore be one which is dominated by more central collisions. The elastic-scattering cross-section is such a process. The integrated cross-section for elastic scattering is

$$\sigma_{\text{el}} = \pi \int db^2 |F(R, R, b, y)|^2 \qquad (8.64)$$

and, since $|F| \sim 1/b^2$, it follows that the elastic cross-section is dominated by more central collisions than the total cross-section.

Having established when we expect the BFKL calculation to break down we turn now to a discussion of the specific nature of the multiple scattering corrections. Firstly we should establish the approximations that are inherent in deriving the particular form of the elastic scattering matrix of the preceding subsection, i.e. Eqs.(8.41) and (8.45). We know that the leading logarithmic approximation of BFKL corresponds to the single scattering of one dipole in one onium with another dipole in the other onium, i.e. the S-matrix for elastic onium–onium scattering can be written

$$S = 1 + \alpha_s^2 \sum_{n=1}^{\infty} c_n (\bar{\alpha}_s y)^n. \qquad (8.65)$$

We have distinguished between factors of α_s and factors of $\bar{\alpha}_s$. The latter factors are always accompanied by a logarithm of the energy since the leading logarithm approximation is also the leading $1/N$ approximation (N being the number of colours). The additional factor of α_s^2 arises due to the colour neutrality of the external onia. In the dipole picture, each onium evolves a dipole cloud by iterating the evolution operator, which is $\sim \bar{\alpha}_s y$. The interaction of the two dipoles is determined by $f \sim \alpha_s^2$. Clearly therefore, the n Pomeron exchange contribution is suppressed by the overall factor $\sim \alpha_s^{2n}$ – so it is sub-leading in both the $1/N$ and leading logarithm approximations. Why, therefore, do we keep these multiple scattering contributions whilst ignoring all the other possible higher order corrections?

The answer is simply stated: it is because of the very high numbers of dipoles which are generated in the evolution of the onia. Typical configurations contain very large numbers of dipoles, i.e. $\sim e^{\omega_0 y'}$ and $\sim e^{\omega_0(y-y')}$, so although the probability of an individual scattering is small ($\sim \alpha_s^2$) the number of 'trials' is very large (it is the product of the number of dipoles in each onium), i.e. $\sim e^{\omega_0 y}$. Thus we expect the multiple scattering corrections to be significant when $\alpha_s^2 e^{\omega_0 y} \sim 1$. The incoherent multiple scattering of the dipoles within the onia, i.e. the exponentiation of the basic dipole–dipole scattering amplitude (e^{-f}), amounts to the assumption that the dominant sub-leading effects are due solely to the large number of dipoles and that collective effects between the individual dipoles (which would spoil the exponentiation) are negligible. To make this plausible consider another sub-leading effect which should become important as the energy increases. This is the effect which we call **dipole saturation**. As the dipole evolution proceeds with the corresponding increase of the dipole number we might expect that dipoles within a single onium start to interact with other dipoles in the same onium. These effects are implemented via a modification of the onium wavefunction and are

$\sim \alpha_s^2 e^{\omega_0 y'}$ (i.e. the amplitude of any given dipole to interact with all the others is proportional to the number of dipoles). Clearly, if we choose to divide the rapidity interval equally between the two onia, i.e. $y' = y/2$, then the saturation effects enter at energies which are roughly the square of the energies where multiple scattering effects first become important. If we choose a highly asymmetric partitioning of the energy (e.g. $y' = 0$ or $y' = y$) then there is no justification for focusing only on the multiple scattering corrections, i.e. there will be large wavefunction saturation effects which alter the evolved onium state in such a way that the total amplitude is the same as that which would be obtained by including only multiple scattering effects but with $y' = y/2$. So, although the physics is clearly independent of y' the sensible choice is $y' = y/2$ since this maximally suppresses the saturation effects which we are unable to calculate. We expect all other sub-leading corrections to be truly sub-leading, i.e. not enhanced by large dipole multiplicity factors.

A word of caution ought to be issued at this stage. The above arguments rely heavily on the fact that the dominant features of the sub-leading corrections can be determined from knowledge of the average features of the dipole evolution. However, one can envisage scenarios where this is a dangerous line of reasoning. For example, consider a collision in the centre of mass (i.e. $y' = y/2$) at very large impact parameters, i.e. in the region where multiple scattering effects are small. One might also infer that saturation effects are therefore even smaller. However, this need not be the case. The dipole evolution could undergo a period of evolution where only small dipoles are produced. These large numbers of localized dipoles may then be subject to significant saturation corrections. In order to contribute to the scattering at large impact parameters at least one large dipole needs to be created (in at least one of the onia) and this may be done at the end of the dipole evolution. Thus the distribution of large dipoles can be affected by what happened earlier in the dipole evolution and hence be subject to large saturation corrections. However, for the typical configurations which provide the dominant contributions to, for example, the total cross-section we expect the more general arguments to hold (Mueller & Salam (1996)).

It is now time to investigate the actual size of the multiple scat-

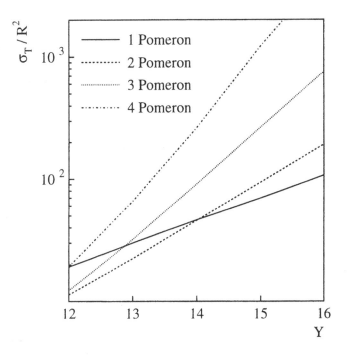

Fig. 8.4. The contributions to the total cross-section for the scattering of two primary dipoles of size R from successive terms in the multiple scattering series (see Salam (1996a)).

tering corrections. The natural course of action is to compute first the corrections to one Pomeron exchange which arise from the $f^2/2!$ term in the exponential series. Progress can be made with an analytic calculation. However, it is not necessary to go into the details here and so we refer to the work of Mueller (1995). The important feature is that the two Pomeron exchange contribution to the onium–onium total cross-section exceeds that for one Pomeron exchange for large enough y. This effect can be seen in Fig. 8.4, where the total cross-section for scattering two primary dipoles each of size R is shown (normalized by R^2). Moreover, and as Fig. 8.4 reveals, the contributions from even more Pomeron exchanges exceed the one Pomeron exchange contribution at

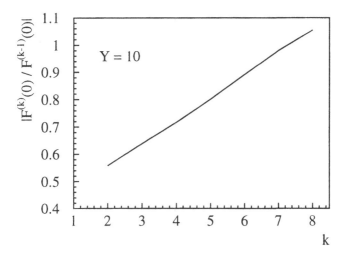

Fig. 8.5. The ratio of the elastic scattering amplitudes for k to $k-1$ Pomeron exchange. The amplitudes are computed at zero impact parameter (see Salam (1996a)).

successively lower energies. The curves are reproduced from the paper by Salam (1996a) using a Monte Carlo program (Salam (1996b)) and with $\alpha_s = 0.18$. Note that the nature of the dipole evolution is ideally suited to the construction of a Monte Carlo program which allows studies far more detailed than are possible analytically.

Some analytic progress has been made in establishing the essential features of the multiple scattering series. In particular, Mueller (1995) has introduced a toy model in which there are no transverse dimensions (i.e. the creation and annihilation operators have no arguments and satisfy $[a, a^\dagger] = 1$). This simplification allows complete analytic calculations to be performed. For large enough energies, the toy model suggests that the terms in the multiple scattering series are $\propto n!$ where n is the number of Pomeron exchanges. This behaviour also seems to hold to a good accuracy in the more realistic QCD case, as Fig. 8.5 shows. The graph shows the ratio of successive terms in the multiple scattering se-

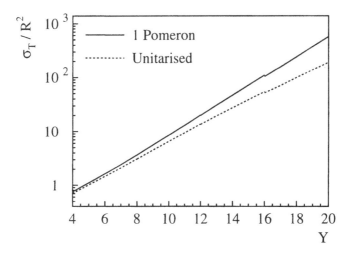

Fig. 8.6. The total cross-section for scattering primary dipoles of size R calculated with and without unitarization corrections (see Salam (1996a)).

ries (for elastic scattering of two primary dipoles at zero impact parameter and fixed energy) and the linearity confirms the factorial behaviour of the terms in the series. This explains the origin of the apparent divergence of the multiple scattering series which is seen in Fig. 8.4. Thus it seems necessary to sum up the whole series before making any predictions. The (summed) large order behaviour of this series cancels out for small enough rapidities even though the individual terms each yield very large contributions. This means that the one Pomeron exchange contribution is good provided the S-matrix is close to unity (as discussed earlier) but as soon as the double Pomeron exchange contribution starts to become important so, too, do all other Pomeron exchanges. The well behaved nature of the fully summed multiple scattering series and the relative smallness of the multiple scattering effects for $y \lesssim 10$ (which is roughly in line with our expectations from the start of this subsection) can be seen in Fig. 8.6, where the total cross-section is shown as a function of rapidity.

To conclude, we have shown how to unitarize the scattering amplitude. Unitarization, via multiple interactions, occurs in the perturbative domain (for small enough primary dipoles). However, non-perturbative physics is ultimately required in order to ensure that the total cross-section satisfies the Froissart–Martin bound (any calculation which assumes massless exchanges, as we do, need not obey that bound). We argued that multiple scattering is the largest unitarization effect. Also, for large enough energies, we argued that the effects of wavefunction saturation can no longer be ignored. Ultimately, the total cross-section becomes dominated by non-perturbative effects. We should also like to remind the reader that we have been working with primary dipoles which are small in size (e.g. dipoles arising from heavy onia). If the colliding particles are light hadrons then the small size configurations are relatively rare fluctuations and lead to small corrections compared with the predominant contribution from non-perturbative physics.

8.3 Summary

• High energy scattering in QCD can be viewed as the scattering of dipole clusters which are generated by the incoherent branching of one dipole into two dipoles. We demonstrated the equivalence of this approach to the one of BFKL developed earlier. The dipole picture provides a very convenient description of high energy scattering in terms of the locations of the dipoles in impact parameter space.

• In QCD the leading logarithm approximation to high energy scattering leads to a power-like growth of total cross-sections, i.e. $\sim s^{\omega_0}$. This growth leads to the violation of unitarity at high enough energies. We quantify when this violation is expected to occur.

• The dipole language of high energy scattering was derived within an operator formalism. This formalism is also suitable for the calculation of the important corrections (to the leading logarithm calculation) which ensure the preservation of unitarity. These corrections arise due to the large number of dipoles within the colliding particles leading to a significant probability that more than one pair of dipoles will interact per onium–onium collision.

8.4 Appendix A

In this appendix we outline the derivation of Eq.(8.7). Our starting point is the relation,

$$\frac{F(s, \mathbf{k_1}, \mathbf{k_2}, \mathbf{q})}{\mathbf{k_2^2}(\mathbf{k_1} - \mathbf{q})^2} = \frac{1}{(2\pi)^2} \int d^2\mathbf{b_{11'}} d^2\mathbf{b_{22'}} d^2(\mathbf{b_1'} - \mathbf{b_2'})$$

$$\times e^{-i(\mathbf{k_1} \cdot \mathbf{b_{11'}} - \mathbf{k_2} \cdot \mathbf{b_{22'}} + \mathbf{q} \cdot (\mathbf{b_1'} - \mathbf{b_2'}))}$$

$$\times (\partial_{\mathbf{b_1}}^2 \partial_{\mathbf{b_1'}}^2)^{-1} \hat{f}(y, \mathbf{b_1}, \mathbf{b_1'}, \mathbf{b_2}, \mathbf{b_2'}), \qquad (A.8.1)$$

which is just the inverse transform of Eq.(4.46).

We now make use of the convolution formula, Eq.(8.6), to replace $\hat{f}$ by a convolution of two $\hat{f}$ factors; after a simple manipulation we find (again using Eq.(4.51) with $n = 0$)

$$\frac{F(s, \mathbf{k_1}, \mathbf{k_2}, \mathbf{q})}{\mathbf{k_2^2}(\mathbf{k_1} - \mathbf{q})^2} = \frac{1}{(2\pi)^2} \int d^2\mathbf{b_{11'}} d^2\mathbf{b_{22'}} d^2(\mathbf{b_1'} - \mathbf{b_2'})$$

$$\times \frac{1}{\pi^8} \int \frac{d^2\mathbf{b_x} d^2\mathbf{b_x'}}{\mathbf{b_{xx'}}^4} d^2\mathbf{c} d^2\mathbf{c'} e^{-i(\cdots)}$$

$$\times \frac{1}{16} \int d\nu \frac{\nu^2}{(\nu^2 + 1/4)^2} \int d\mu\, \mu^2 e^{\bar{\alpha}_s(\chi_0(\nu)y' + \chi_0(\mu)(y - y'))}$$

$$\times \tilde{\phi}_0^\nu(\mathbf{b_1}, \mathbf{b_1'}, \mathbf{c}) \tilde{\phi}_0^{\nu*}(\mathbf{b_x}, \mathbf{b_x'}, \mathbf{c}) \tilde{\phi}_0^\mu(\mathbf{b_x}, \mathbf{b_x'}, \mathbf{c'}) \tilde{\phi}_0^{\mu*}(\mathbf{b_2}, \mathbf{b_2'}, \mathbf{c'}), (A.8.2)$$

where $\mathbf{b_{xx'}} = \mathbf{b_x} - \mathbf{b_{x'}}$.

Now we insert the delta function operator:

$$\delta^2(\mathbf{b_x} - \mathbf{b_y})\delta^2(\mathbf{b_x'} - \mathbf{b_y'}) + \delta^2(\mathbf{b_x} - \mathbf{b_y'})\delta^2(\mathbf{b_x'} - \mathbf{b_y})$$

$$= \frac{1}{2(2\pi)^4} \partial_{\mathbf{b_y}}^2 \partial_{\mathbf{b_y'}}^2 \int \frac{d^2\mathbf{l_1}}{\mathbf{l_1}^2} \frac{d^2\mathbf{l_2}}{\mathbf{l_2}^2}$$

$$\times \left[e^{i\mathbf{l_1} \cdot (\mathbf{b_x} - \mathbf{b_y})} - e^{i\mathbf{l_1} \cdot (\mathbf{b_x} - \mathbf{b_y'})} - e^{i\mathbf{l_1} \cdot (\mathbf{b_x'} - \mathbf{b_y})} + e^{i\mathbf{l_1} \cdot (\mathbf{b_x'} - \mathbf{b_y'})} \right]$$

$$\times \left[e^{i\mathbf{l_2} \cdot (\mathbf{b_x} - \mathbf{b_y})} - e^{i\mathbf{l_2} \cdot (\mathbf{b_x} - \mathbf{b_y'})} - e^{i\mathbf{l_2} \cdot (\mathbf{b_x'} - \mathbf{b_y})} + e^{i\mathbf{l_2} \cdot (\mathbf{b_x'} - \mathbf{b_y'})} \right].$$

$$(A.8.3)$$

Since the eigenfunctions, $\tilde{\phi}_0^\mu(\mathbf{b_1}, \mathbf{b_2}, \mathbf{c})$, are symmetric under the interchange of the first two arguments (i.e. $\mathbf{b_1} \leftrightarrow \mathbf{b_2}$) we can insert the delta functions of Eq.(A.8.3). Note that the $\mathbf{l_i}$ integrals are finite *before* the action of the Laplacian operators — we utilized the symmetry property of the eigenfunctions under interchange of the arguments to ensure just this property. As a result, we can perform an integration by parts to reverse the action of the Laplacian operators such that they act upon the eigenfunctions rather than the l integrand. Hence,

$$\frac{F(s, \mathbf{k_1}, \mathbf{k_2}, \mathbf{q})}{\mathbf{k_2^2}(\mathbf{k_1} - \mathbf{q})^2} = \frac{1}{4(2\pi)^6} \int d^2\mathbf{b_{11'}} d^2\mathbf{b_{22'}} d^2(\mathbf{b_1'} - \mathbf{b_2'}) e^{-i(\cdots)}$$

$$\times \int \frac{d^2\mathbf{b_x} d^2\mathbf{b_x'}}{\mathbf{b_{xx'}}^4} \frac{d^2\mathbf{b_y} d^2\mathbf{b_y'}}{\mathbf{b_{yy'}}^4} d^2\mathbf{c} d^2\mathbf{c'} \frac{d^2\mathbf{l_1}}{\mathbf{l_1}^2} \frac{d^2\mathbf{l_2}}{\mathbf{l_2}^2} [\cdots][\cdots]$$

$$\times \frac{1}{\pi^8} \int d\nu\, \nu^2 \int d\mu\, \mu^2 e^{\bar{\alpha}_s(\chi_0(\nu)y' + \chi_0(\mu)(y-y'))}$$

$$\times \tilde{\phi}_0^\nu(\mathbf{b_1}, \mathbf{b_1'}, \mathbf{c}) \tilde{\phi}_0^{\nu*}(\mathbf{b_y}, \mathbf{b_y'}, \mathbf{c}) \tilde{\phi}_0^\mu(\mathbf{b_x}, \mathbf{b_x'}, \mathbf{c'}) \tilde{\phi}_0^{\mu*}(\mathbf{b_2}, \mathbf{b_2'}, \mathbf{c'}). \quad (A.8.4)$$

Making the standard change of variables

$$\mathbf{R_x} = \frac{1}{2} (\mathbf{b_x} + \mathbf{b_x'}) - \mathbf{c'},$$

and similarly for the other co-ordinates, allows the volume elements to be re-written, i.e.

$$d^2(\mathbf{b_1'} - \mathbf{b_2'}) d^2\mathbf{c} d^2\mathbf{c'} \rightarrow d^2(\mathbf{c} - \mathbf{c'}) d^2\mathbf{R_1} d^2\mathbf{R_2}$$

$$d^2\mathbf{b_x} d^2\mathbf{b_x'} \rightarrow d^2\mathbf{R_x} d^2\mathbf{b_{xx'}}, \qquad \text{etc.} \quad (A.8.5)$$

The independent variables are now $\mathbf{b_{11'}}, \mathbf{b_{22'}}, \mathbf{R_1}, \mathbf{R_2}, \mathbf{b_{xx'}}, \mathbf{b_{yy'}},$ $\mathbf{R_x}, \mathbf{R_y}$ and $\mathbf{c} - \mathbf{c'}$. The only dependence upon $\mathbf{c} - \mathbf{c'}$ is in the exponential terms. Hence we can collect them together and integrate over $\mathbf{c} - \mathbf{c'}$ which gives the delta function factor $(2\pi)^2 \delta^2(\mathbf{q} - \mathbf{l_1} - \mathbf{l_2})$.

The remaining integrals, combined with the definitions specified by Eqs.(8.8) and (8.9), lead directly to the desired result, i.e. Eq.(8.7).

8.5 Appendix B

In this appendix we derive Eq.(8.27) for the probability of emission of a gluon from a dipole.

Consider a colour singlet dipole with momentum p_1 moving along the positive z-axis. It is convenient to define a momentum p_2 with the same energy component moving along the negative z-axis, such that $2p_1 \cdot p_2 = s$.

Let $\psi(\rho_0, \mathbf{r})$ be the amplitude for this dipole to consist of a quark–antiquark pair in which the the quark carries a fraction ρ_0

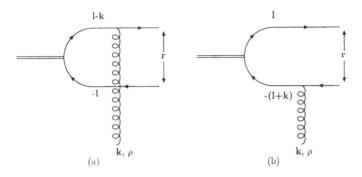

Fig. 8.7. Graphs for the emission of a gluon from a dipole.

of the longitudinal momentum of the dipole and is separated from
the antiquark, in impact parameter space, by **r**. We may write this
in terms of an amplitude in transverse momentum space as

$$\psi(\rho_0, \mathbf{r}) = \frac{1}{(2\pi)^2} \int d^2 \mathbf{l}\, e^{i\mathbf{l}\cdot\mathbf{r}}\, \tilde{\psi}(\rho_0, \mathbf{l}), \qquad (B.8.1)$$

where $\tilde{\psi}(\rho_0, \mathbf{l})$ is the amplitude for the quark to have transverse
momentum **l** and the antiquark to have transverse momentum $-\mathbf{l}$.

Now consider a gluon with transverse momentum **k** and fraction
of longitudinal momentum ρ emitted from this dipole, as shown in
Fig. 8.7. We assume that $\rho \ll \rho_0,\ (1-\rho_0)$. This is the strong ordering required for the leading logarithm approximation. This gluon
will later couple to a further gluon so it is really off-shell. However,
since any gluon to which it couples has a fraction of longitudinal
momentum which is small compared with that of the parent gluon
and transverse momentum which is small compared with the longitudinal momentum of the parent, it is a valid approximation to
consider the emitted gluon to be on shell (and hence transversely
polarized). The rapidity of the emitted gluon is given by

$$y = \frac{1}{2}\ln\left(s\rho^2/\mathbf{k}^2\right). \qquad (B.8.2)$$

We may write the momenta of the quark and antiquark as

$$l_1^\mu = \rho_0 p_1^\mu + \frac{\mathbf{l}^2}{s\rho_0} p_2^\mu + l_\perp^\mu, \qquad (B.8.3)$$

$$l_2^\mu = (1 - \rho_0)p_1^\mu + \frac{\mathbf{l}^2}{s(1 - \rho_0)}p_2^\mu - l_\perp^\mu, \qquad (\text{B.8.4})$$

and the momentum of the emitted gluon as

$$k^\mu = \rho p_1^\mu + \frac{\mathbf{k}^2}{s\rho}p_2^\mu + k_\perp^\mu. \qquad (\text{B.8.5})$$

Furthermore we can exploit gauge invariance to demand that the polarization vector, e^μ, of the emitted gluon has no component proportional to p_1^μ and, using the fact that the gluon is transverse $(e \cdot k = 0)$, we have

$$e^\mu = \frac{2\mathbf{e} \cdot \mathbf{k}}{s\rho}p_2^\mu + e_\perp^\mu. \qquad (\text{B.8.6})$$

Now the amplitude for emission from the quark (Fig. 8.7(a)) is

$$- ig\tau^a \frac{2l_1 \cdot e}{2l_1 \cdot k}\tilde\psi(\rho_0, \mathbf{l} - \mathbf{k}/2), \qquad (\text{B.8.7})$$

where the factor $\tilde\psi(\rho_0, \mathbf{l} - \mathbf{k}/2)$ indicates that the quark–antiquark pair produced by the dipole are separated by $2\mathbf{l} - \mathbf{k}$ in transverse momentum space. τ^a is the colour generator in the fundamental representation. We have used the eikonal approximation as the emitted gluon is soft relative to the parent quark.

For $\rho \ll \rho_0$, we may use Eqs.(B.8.3), (B.8.5) and (B.8.6) to write

$$2l_1 \cdot e = \frac{2\rho_0\mathbf{e} \cdot \mathbf{k}}{\rho}$$

$$2l_1 \cdot k = \frac{\rho_0}{\rho}\mathbf{k}^2$$

(we have kept only the terms proportional to $1/\rho$), so that the contribution from this graph becomes

$$- 2ig\tau^a\frac{\mathbf{e} \cdot \mathbf{k}}{\mathbf{k}^2}\tilde\psi(\rho_0, \mathbf{l} - \mathbf{k}/2). \qquad (\text{B.8.8})$$

Likewise the contribution from emission off the antiquark (Fig. 8.7(b)) is

$$2ig\tau^a\frac{\mathbf{e} \cdot \mathbf{k}}{\mathbf{k}^2}\tilde\psi(\rho_0, \mathbf{l} + \mathbf{k}/2). \qquad (\text{B.8.9})$$

Returning to impact parameter space the total amplitude for the emission of the gluon off a dipole with transverse size $\mathbf{r}$ is

$$-2ig\tau^a \frac{\mathbf{e} \cdot \mathbf{k}}{\mathbf{k}^2} \frac{1}{(2\pi)^2} \int d^2l e^{i\mathbf{l} \cdot \mathbf{r}} \left(\tilde{\psi}(\rho_0, \mathbf{l} - \frac{\mathbf{k}}{2}) - \tilde{\psi}(\rho_0, \mathbf{l} + \frac{\mathbf{k}}{2}) \right)$$

$$= -2ig\tau^a \frac{\mathbf{e} \cdot \mathbf{k}}{\mathbf{k}^2} e^{-i\frac{1}{2}\mathbf{k} \cdot \mathbf{r}} \frac{1}{(2\pi)^2} \int d^2l e^{i\mathbf{l} \cdot \mathbf{r}} \tilde{\psi}(\rho_0, \mathbf{l}) \left(1 - e^{i\mathbf{k} \cdot \mathbf{r}} \right)$$

$$= -2ig\tau^a \frac{\mathbf{e} \cdot \mathbf{k}}{\mathbf{k}^2} e^{-i\frac{1}{2}\mathbf{k} \cdot \mathbf{r}} \psi(\rho_0, \mathbf{r}) \left(1 - e^{i\mathbf{k} \cdot \mathbf{r}} \right). \qquad \text{(B.8.10)}$$

Taking the square modulus of this and summing over emitted gluon polarizations and colours gives us

$$4g^2 \frac{N^2 - 1}{N} \frac{|\psi(\rho_0, \mathbf{r})|^2}{\mathbf{k}^2} \left(1 - e^{i\mathbf{k} \cdot \mathbf{r}} \right) \qquad \text{(B.8.11)}$$

(where we understand that we must take the real part of the exponential). The factor of $(N^2 - 1)/2N$ comes from the square of the colour generator summed over all possible colours for the emitted gluon. In the large N limit we may replace this by $N/2$ (this allows us to generalize our result without modification, so that it describes gluon emission off any colour dipole, i.e. not just a q–$\bar{q}$ pair). It is worth noting here that this expression is proportional to the impact factor for the coupling of a (zero momentum transfer) Pomeron to the parent dipole. The formalism is easily extended to non-zero momentum transfer.[†]

The element of phase space is given by

$$\frac{1}{2(2\pi)^3} d^2\mathbf{k} \frac{d\rho}{\rho}.$$

We may use Eq.(B.8.2) to express this in terms of rapidity and obtain

$$\frac{1}{2(2\pi)^3} d^2\mathbf{k} dy.$$

Thus the probability of emitting a gluon into a rapidity interval dy and transverse momentum interval $d^2\mathbf{k}$ is

[†] For non-zero momentum transfer, we need to multiply Eq.(B.8.10) by the conjugate of the amplitude which is obtained by replacing $\mathbf{k} \to \mathbf{k} - \mathbf{q}$ in Eq.(B.8.10). This adds extra exponential factors in the final result as well a factor of $(\mathbf{e} \cdot \mathbf{k})(\mathbf{e} \cdot (\mathbf{k} - \mathbf{q}))$. The latter factor poses no problem, on using Eq.(8.29), whilst the former (on transforming to impact parameter space) leads to delta functions which fix the locations of the dipoles (i.e. the arguments of the creation and annihilation operators of Eq.(8.32)).

$$\frac{\bar{\alpha}_s}{\pi} \frac{d^2\mathbf{k}}{\mathbf{k}^2} dy \left(1 - e^{i\mathbf{k}\cdot\mathbf{r}}\right) |\psi(\rho_0, \mathbf{r})|^2. \qquad \text{(B.8.12)}$$

The factor $|\psi(\rho_0, \mathbf{r})|^2$ is just the probability of finding the dipole in the first place. Therefore the probability for emission of a gluon from a dipole is given by Eq.(8.27).

Appendix: Feynman rules for QCD

Propagators:

Gluon propagator (Feynman gauge):

$\mu\overset{k}{}\nu$
ab $-i\delta_{ab}g_{\mu\nu}/(k^2 + i\epsilon)$

(Massless) fermion propagator:

$\beta\overset{k}{}\alpha$
ij $i\delta_{ij}(\gamma \cdot k)_{\alpha\beta}/(k^2 + i\epsilon)$

(Massless) scalar propagator:

$\overset{k}{}$
$\bar{i}\bar{j}$ $i\delta_{ij}/(k^2 + i\epsilon)$

Vertices: (momenta are always outgoing)

Scalar–gluon interactions:

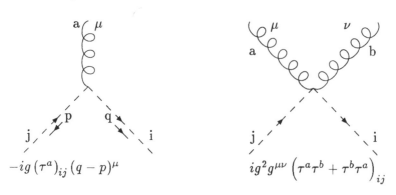

$-ig\,(\tau^a)_{ij}\,(q-p)^\mu$ $ig^2 g^{\mu\nu}\left(\tau^a\tau^b + \tau^b\tau^a\right)_{ij}$

Fermion–gluon interaction:

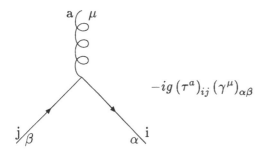

$$-ig\,(\tau^a)_{ij}\,(\gamma^\mu)_{\alpha\beta}$$

Gluon self-interactions:

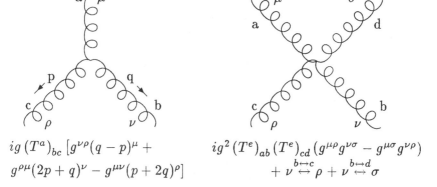

$ig\,(T^a)_{bc}\,[g^{\nu\rho}(q-p)^\mu +$
$g^{\rho\mu}(2p+q)^\nu - g^{\mu\nu}(p+2q)^\rho]$

$ig^2\,(T^e)_{ab}\,(T^e)_{cd}\,(g^{\mu\rho}g^{\nu\sigma} - g^{\mu\sigma}g^{\nu\rho})$
$+\ \nu \overset{b\leftrightarrow c}{\leftrightarrow} \rho + \nu \overset{b\leftrightarrow d}{\leftrightarrow} \sigma$

The gauge coupling constant is g $(\alpha_s = g^2/4\pi)$.

The matrices $(\tau^a)_{ij}$ are the matrices of the colour group in the representation of the quarks or colour scalar particles.

The matrices $(T^a)_{bc} = -if_{abc}$ are the colour matrices in the adjoint representation and f_{abc} are the structure constants of the colour group.

In addition there is a further factor of $-i$ accompanying each amplitude so that it is an element of the T-matrix rather than the S-matrix $(S = \mathbb{1} + iT)$.

References

Abachi, S. *et al.* (1994) *Phys. Rev. Lett.* **72**, 2332

Abatzis, S. *et al.* (1994) *Phys. Lett.* **B324**, 509

Abe, F. *et al.* (1995) *Phys. Rev. Lett.* **74**, 855

Abers, E. & Teplitz, V.L. (1967) *Phys. Rev.* **158**, 1365

Abers, E., Keller, R.A. & Teplitz, V.L. (1970) *Phys. Rev.* **D2**, 1757

Abramowicz, H., Frankfurt, L. & Strikman, M. (1995) *DESY preprint* **95-047** hep-ph/9503437

Abramowitz, M. & Stegun, I.A. (1972) *Handbook of Mathematical Functions*, Dover Publications Inc., New York

Ahmed, T. *et al.* (1994) *Nucl. Phys.* **B429** 477

Ahmed, T. *et al.* (1995a) *Nucl. Phys.* **B435** 3

Ahmed, T. *et al.* (1995b) *Phys. Lett.* **B348** 681

Ahmed, T. *et al.* (1995c) *Nucl. Phys.* **B439** 471

Altarelli, G. & Parisi, G. (1977) *Nucl. Phys.* **B126**, 298

Amos, N. *et al.* (1992) *Phys. Rev. Lett.* **68**, 2433

Augier, C. *et al.* (1993) *Phys. Lett.* **B316**, 448

Balitsky, Y.Y. & Lipatov, L.N. (1978) *Sov. J. Nucl. Phys.* **28**, 822

Barnes, A.V. *et al.* (1976) *Phys. Rev. Lett.* **37**, 76

Bartels, J. (1975) *Nucl. Phys.* **B151**, 293

Bartels, J. (1980) *Nucl. Phys.* **B175**, 365

Bartels, J. (1993a) *Zeit. Phys.* **C60**, 471

Bartels, J. (1993b) *Phys. Lett.* **B298**, 204

Bartels, J. *et al.* (1995) *Phys. Lett.* **B348**, 589

Bartels, J., De Roeck, A. & Loewe, M. (1992) *Zeit. Phys.* **C54**, 635

Bartels, J., Forshaw, J.R., Lotter, H. & Wüsthoff, M., (1996) *Phys. Lett.* **B375**, 301

Bartels, J. & Lotter, H. (1993) *Phys. Lett.* **B309**, 400

Bartels, J. & Wüsthoff, M. (1995) *Zeit. Phys.* **C66**, 157

Bartels, J., Wüsthoff, M. & Lipatov, L.N. (1995) *Nucl. Phys.* **B464**, 298

Bassetto, A., Ciafaloni, M. & Marchesini, G. (1983) *Phys. Rep.* **100**, 201

Bjorken, J.D. (1994) *SLAC preprint* **SLAC-PUB-6477**

Bouquet, A., Diu, B., Leader, E. & Nicolescu, B. (1975) *Nuov. Cim.* **29A**, 30

Brodsky, S.J., *et al.* (1994) *Phys. Rev.* **D50** 3134

Büttner, K. & Pennington, M.R. (1995) *Phys. Lett.* **B356** 354.

Carlson, F. (1914) *Upsala thesis*

Catani, S., Ciafaloni, M. & Hautmann, F. (1990) *Phys. Lett.* **B242**, 97

Catani, S., Ciafaloni, M. & Hautmann, F. (1991) *Nucl. Phys.* **B366**, 135

Catani, S., Fiorani, F. & Marchesini, G. (1990a) *Phys. Lett.* **B234**, 339

Catani, S., Fiorani, F. & Marchesini, G. (1990b) *Nucl. Phys.* **B336**, 18

Catani, S., Fiorani, F. Marchesini, G. & Oriani, G. (1991) *Nucl. Phys.* **B361**, 645

Catani, S. & Hautmann, F. (1994) *Nucl. Phys.* **B427** 475

Chang, S.J. & Yan, T.M. (1970)*Phys. Rev. Lett.* **25**, 1586

Chang, S.J. & Yan, T.M. (1971) *Phys. Rev.* **D4**, 537

Chen, Z. & Mueller, A.H. (1995) *Nucl. Phys.* **B451**, 579

Cheng, H. & Lo, C.Y. (1976) *Phys. Rev.* **D13**, 1131

Cheng, H. & Lo, C.Y. (1977) *Phys. Rev.* **D15**, 2959

Cheng, H. & Wu, T.T. (1965) *Phys. Rev.* **140B**, 465

Cheng, H. & Wu, T.T. (1969a) *Phys. Rev.* **182**, 1852

Cheng, H. & Wu, T.T. (1969b) *Phys. Rev.* **182**, 1868

Cheng, H. & Wu, T.T. (1969c) *Phys. Rev.* **182**, 1873

Cheng, H. & Wu, T.T. (1970a) *Phys. Rev. Lett.* **124**, 1456

Cheng, H. & Wu, T.T. (1970b) *Phys. Rev.* **D1**, 1069

Cheng, H. & Wu, T.T. (1970c) *Phys. Rev.* **D1**, 1083

Chew, G.F.P. & Frautschi, S.C. (1961) *Phys. Rev. Lett.* **7**, 394

Chew, G.F.P. & Frautschi, S.C. (1962) *Phys. Rev. Lett.* **8**, 41

Ciafaloni, M. (1988) *Nucl. Phys.* **B296**, 49

Close, F.E. (1979) *An Introduction to Quarks and Partons*, Academic Press, London

Collins, J.C. & Ellis, R.K. (1991) *Nucl. Phys.* **B360**, 3

Collins, J.C. & Kwiecinski, J. (1989) *Nucl. Phys.* **B316**, 307

Collins, J.C. & Landshoff, P.V. (1992) *Phys. Lett.* **B276**, 196

Collins, J.C., Soper, D.E. & Sterman, G. (1989) in *Perturbative QCD*, ed. A.H. Mueller, World Scientific, Singapore

Collins, P.D.B. (1977) *Introduction to Regge Theory and High Energy Physics*, Cambridge University Press

Corianò, C. & White, A.R. (1995) *Acta Physica Polonica* **B26**, 2005

Corianò, C. & White, A.R. (1996) *Nucl. Phys.* **B468**, 175

Cutkosky, R.E. (1960) *J. Math. Phys.* **1**, 429

Daniell, G.J. & Ross, D.A. (1989) *Phys. Lett.* **B224**, 166

Del Duca, V. (1995) *Scientifica Acta* **Vol. 10**, 91

Del Duca, V. & Schmidt, C.R. (1994a) *Phys. Rev.* **D49**, 177

Del Duca, V. & Schmidt, C.R. (1994b) *Phys. Rev.* **D49**, 4510

Derrick, M. *et al.* (1993) *Phys. Lett.* **B315**, 481

Derrick, M. *et al.* (1995a) *Phys. Lett.* **B346**, 399

Derrick, M. *et al.* (1995b) *Phys. Lett.* **B356**, 129

Derrick, M. *et al.* (1995c) *Zeit. Phys.* **C65**, 379

Derrick, M. *et al.* (1996a) *Zeit. Phys.* **C70**, 391

Derrick, M. *et al.* (1996b) *Phys. Lett.* **B369**, 55

Dicus, D.A. & Teplitz, V.L. (1971) *Phys. Rev.* **D3**, 1910

Dokshitzer, Yu.L. (1977) *Sov. Phys. JETP* **73**, 1216

Dokshitzer, Yu.L., Khoze, V.A., Troyan, S.I. & Mueller, A.H. (1991) *Basics of Perturbative QCD*, Editions Frontières, Paris.

Donnachie, A. & Landshoff, P.V. (1984) *Nucl. Phys.* **B244**, 322

Donnachie, A. & Landshoff, P.V. (1992) *Phys. Lett.* **B296**, 227

Eden, R.J., Landshoff, P.V., Olive, D.I. & Polkinghorne J.C. (1966) *The Analytic S-Matrix*, Cambridge University Press

Faddeev, L.D. & Korchemsky, G.P. (1995) *Phys. Lett.* **B342**, 311

Fadin, V.S. & Fiore, R. (1992) *Phys. Lett.* **B294**, 286

Fadin, V.S., Fiore, R. & Kotsky, M.I. (1996) *Phys. Lett.* **B387**, 593

Fadin, V.S., Fiore, R. & Quartarolo, A. (1994a) *Phys. Rev.* **D50**, 2265

Fadin, V.S., Fiore, R. & Quartarolo, A. (1994b) *Phys. Rev.* **D50**, 5893

Fadin, V.S., Fiore, R. & Quartarolo, A. (1996) *Phys. Rev.* **D53**, 2729

Fadin, V.S., Kuraev E.A. & Lipatov, L.N. (1975) *Phys. Lett.* **B60**, 50

Fadin, V.S., Kuraev E.A. & Lipatov, L.N. (1976) *Sov. Phys. JETP* **44**, 443

Fadin, V.S., Kuraev E.A. & Lipatov, L.N. (1977) *Sov. Phys. JETP* **45**, 199

Fadin, V.S. & Lipatov, L.N. (1992) *Nucl. Phys.* B (Proc. Suppl.) **29A** 93

Fadin, V.S. & Lipatov, L.N. (1993) *Nucl. Phys.* **B406**, 259

Fadin, V.S. & Lipatov, L.N. (1996) *Nucl. Phys.* **B477**, 767

Field, R.D. (1989) *Applications of Perturbative QCD*, Addison-Wesley

Foldy, L.F. & Peierls, R.F. (1963) *Phys. Rev.* **130**, 1585

Forshaw, J.R. & Ryskin, M.G. (1995) *Zeit. Phys.* **C68**, 137

Frankfurt, L.L. & Sherman, V.E. (1976) *Sov. J. Nucl. Phys.* **23**, 581

Froissart, M. (1961)*Phys. Rev.* **123**, 1053

Frolov G.V. & Lipatov L.N. (1971) *Sov. J. Nucl. Phys.* **13**, 333

Frolov G.V., Gribov, V.N. & Lipatov, L.N. (1970) *Phys. Lett.* **B31**, 34

Frolov G.V., Gribov, V.N. & Lipatov, L.N. (1971) *Sov. J. Nucl. Phys.* **12**, 543

Gell-Mann, M. & Goldberger, M.L. (1962) *Phys. Rev. Lett.* **9**, 275

Gell-Mann, M., Goldberger, M.L., Low, F.E., Marx, E. & Zachariasen, F. (1964a) *Phys. Rev.* **133B**, 145

Gell-Mann, M., Goldberger, M.L., Low, F.E., Singh, V. & Zachariasen, F. (1964b) *Phys. Rev.* **133B**, 161

Good, M.L. & Walker, W.D. (1960) *Phys. Rev.* **120**, 1857

Gradshteyn, I.S. & Ryzhik, I.M. (1994), *Tables of Integrals, Series, and Products*, 5th Edition, Academic Press

Greiner, W. & Schäfer, A. (1994) *Quantum Chromodynamics*, Springer

Gribov, V.N. (1968) *Sov. Phys. JETP* **26**, 414

Gribov, V.N. & Lipatov, L.N. (1972) *Sov. J. Nucl. Phys.* **15** 78

Gribov, L.V., Levin, E.M. & Ryskin, M.G. (1983) *Phys. Rep.* **100**, 1

Grisaru, M. T., Schnitzer, H.J. & Tsao H.S. (1973) *Phys. Rev.* **D8**, 4498

Halzen, F. & Martin, A.D. (1984) *Quarks & Leptons: An Introductory Course in Modern Particle Physics*, John Wiley & Sons

Hancock, R.E. & Ross, D.A. (1992) *Nucl. Phys.* **B383**, 575

Hancock, R.E. & Ross, D.A. (1993) *Nucl. Phys.* **B394**, 200

Ingelman, G. & Schlein, P. (1985) *Phys. Lett.* **B152**, 256

Jaroszewicz, T. (1980) *Acta Physica Polonica* **B11**, 965

Joynson, D., Leader, E., Lopez, C. & Nicolescu, B. (1975) *Nuov. Cim.* **30A**, 345

Kinoshita, T. (1962) *J. Math. Phys.* **3**, 650

Kirschner, R. (1994) *Zeit. Phys.* **C65** 505

Kirschner, R. & Lipatov, L.N. (1990) *Zeit. Phys.* **C45**, 477

Kirschner, R., Lipatov, L.N. & Szymanowski, L. (1994) *Nucl. Phys.* **B425**, 579

Korchemsky, G.P. (1995) *Nucl. Phys.* **B443** 255

Korchemsky, G.P. (1996) *Nucl. Phys.* **B462**, 333

Kwiecinski, J., Martin, A.D. & Sutton, P.J. (1992) *Phys. Rev.* **D46** 971

Kwiecinski, J., Martin, A.D. & Sutton, P.J. (1995) *Phys. Rev.* **D52** 1445

Kwiecinski, J. & Praszalowicz, M. (1980) *Phys. Lett.* **B94** 413

Landshoff, P.V. (1974) *Phys. Rev.* **D10**, 1024

Landshoff, P.V. & Nachtmann, O. (1987) *Zeit. Phys.* **C35**, 405

Landshoff, P.V., Polkinghorne, J.C. & Short, R.D. (1970) *Nucl. Phys.* **B28**, 210

Landshoff, P.V. & Polkinghorne, J.C. (1971) *Nucl. Phys.* **B32**, 541

Lee, T.D. & Nauenberg, M. (1964) *Phys. Rev.* **B133**, 1549

Levin, E.M., Ryskin, M.G., Shabelskii, Yu.M. & Shuvaev, A.G. (1991) *Sov. J. Nucl. Phys.* **53**, 657

Levin, E.M. & Tan, C. (1992) *Brown preprint* **HET-889** hep-ph/9302308

Lipatov, L.N. (1976) *Sov. J. Nucl. Phys.* **23**, 338

Lipatov, L.N. (1986) *Sov. Phys. JETP* **63**, 904

Lipatov, L.N. (1989) in *Perturbative Quantum Chromodynamics*, Ed. A.H. Mueller, World Scientific, Singapore

Lipatov, L.N. (1990) *Phys. Lett.* **B251**, 284

Lipatov, L.N. (1991) *Nucl. Phys.* **B365**, 641

Lipatov, L.N. (1993) *Phys. Lett.* **B309**, 394

Lipatov, L.N. (1994) *JETP Lett.* **59**, 596

Lipatov, L.N. (1995) *Nucl. Phys.* **B452** 369

Lipatov, L.N. & Fadin, V.S. (1989a) *Sov. J. Nucl. Phys.* **50** 712

Lipatov, L.N. & Fadin, V.S. (1989b) *JETP Lett.* **49** 352

Low, F.E. (1975) *Phys. Rev.* **D 12**, 163

Mandelstam, S. (1965) *Phys. Rev.* **137B**, 949

Marchesini, G. (1995) *Nucl. Phys.* **B445**, 49

Martin, A. (1963) *Phys. Rev* **129**, 1432

Martin, A.D., Kwiecinski, J. & Sutton, P.J. (1992) *Nucl. Phys.* B (proc. supp.) **29A** 67.

Martin, B.R., Morgan, D. & Shaw, G. (1976) *Pion-Pion Interactions in Particle Physics*, Academic Press

Mason, A.L. (1976a) *Nucl. Phys.* **B104**, 141

Mason, A.L. (1976b) *Nucl. Phys.* **B117**, 493

Mason, A.L. (1977) *Nucl. Phys.* **B120**, 285

Mathews, J. & Walker, R.L. (1970) *Mathematical Methods of Physics*, W.A. Benjamin, New York

McCoy, B.M. & Wu, T.T. (1976a) *Phys. Rev.* **13**, 369

McCoy, B.M. & Wu, T.T. (1976b) *Phys. Rev.* **13**, 379

McCoy, B.M. & Wu, T.T. (1976c) *Phys. Rev.* **13**, 395

McCoy, B.M. & Wu, T.T. (1976d) *Phys. Rev.* **13**, 424

McCoy, B.M. & Wu, T.T. (1976e) *Phys. Rev.* **13**, 484

McCoy, B.M. & Wu, T.T. (1976f) *Phys. Rev.* **13**, 508

McDermott, M.F., Forshaw, J.R. & Ross, G.G. (1995) *Phys. Lett.* **B349**, 189

Miettinen, H.I. & Pumplin, J. (1978) *Phys. Rev.* **D18**, 1696

Mueller, A.H. (1991) *Nucl. Phys.* B (proc. supp.) **18C**, 125

Mueller, A.H. (1994) *Nucl. Phys.* **B415**, 373

Mueller, A.H. (1995) *Nucl. Phys.* **B437**, 107

Mueller, A.H. & Navelet, H. (1987) *Nucl. Phys.* **B282**, 727

Mueller, A.H. & Patel, B. (1994) *Nucl. Phys.* **B425**, 471

Mueller, A.H. & Qiu, J. (1986) *Nucl. Phys.* **B268**, 427

Mueller, A.H. & Salam, G.P. (1996) *Nucl. Phys.* **B475**, 293

Mueller, A.H. & Tang, W.-K. (1992) *Phys. Lett.* **B284**, 123

Nikolaev, N.N. (1981) *Sov. Phys. JETP* **54**, 434

Nikolaev, N.N. & Zakharov, B.G. (1994) *Zeit. Phys.* **C64**, 631
Nikolaev, N.N., Zakharov, B.G. & Zoller, V.R. (1994a) *JETP Lett.* **60**, 694
Nikolaev, N.N., Zakharov, B.G. & Zoller, V.R. (1994b) *Phys. Lett.* **B328**, 486
Nussinov, S. (1975) *Phys. Rev. Lett.* **34**, 163
Nussinov, S. (1976) *Phys. Rev.* **D4**, 246
Okun, L.B. & Pomeranchuk, I. Y. (1956) *Sov. Phys. JETP* **3**, 307
Polkinghorne, J.C. (1963a) *J. Math. Phys.* **4**, 503
Polkinghorne, J.C. (1963b) *J. Math. Phys.* **4**, 1393
Polkinghorne, J.C. (1963c) *J. Math. Phys.* **4**, 1396
Polkinghorne J.C. (1964) *J. Math. Phys.* **5**, 1491
Pomeranchuk, I.Y. (1956) *Sov. Phys.* **3**, 306
Pomeranchuk, I.Y. (1958) *Sov. Phys.* **7**, 499
Regge, T. (1959) *Nuov. Cim.* **14**, 951
Regge, T. (1960) *Nuov. Cim.* **18**, 947
Roberts, R.G. (1990) *The Structure of the Proton*, Cambridge University Press
Ryskin, M.G. (1993) *Zeit. Phys.* **C37** 89
Salam, G.P. (1996a) *Nucl. Phys.* **B461** 512
Salam, G.P. (1996b) *Cavendish-HEP preprint* **95/07** hep-ph/9601220
Sen, A. (1983) *Phys. Rev.* **D27**, 2997
Schlein, P. (1993) *Nucl. Phys. B* (proc. supp.) **33A**, 41
Sommerfeld, A. (1949) *Partial Differential Equations in Physics*, Academic Press
Stirling, W.J. (1994) *Nucl. Phys.* **B423**, 56
Tang, W.-K. (1992) *Phys. Lett.* **B278** 363.
Tyburski, L. (1976) *Phys. Rev.* **D13**, 1107
Verlinde, H. & Verlinde, E. (1993) *PUPT preprint* **1319** hep-th/9302104
Watson, G.N. (1918) *Proc. Roy. Soc.* **95**, 83

Index

Printed in the United States
By Bookmasters